Greenhouse Technology

Greenhouse Technology
Principle and Practices

Arupratan Ghosh

CRC Press
Taylor & Francis Group
Boca Raton London New York

CRC Press is an imprint of the
Taylor & Francis Group, an **informa** business

NEW INDIA PUBLISHING AGENCY
New Delhi – 110 034

CRC Press
Taylor & Francis Group
6000 Broken Sound Parkway NW, Suite 300
Boca Raton, FL 33487-2742

First issued in paperback 2023

ISBN-13: 978-0-367-46238-3 (hbk)
ISBN-13: 978-1-03-265449-2 (pbk)

Print edition not for sale in South Asia (India, Sri Lanka, Nepal, Bangladesh, Pakistan or Bhutan)

Publisher's Note
The publisher has gone to great lengths to ensure the quality of this reprint but points out that some imperfections in the original copies may be apparent.

Library of Congress Cataloging-in-Publication Data
A catalog record has been requested

Visit the Taylor & Francis Web site at
http://www.taylorandfrancis.com

and the CRC Press Web site at
http://www.crcpress.com

Preface

Since last 100 years only the Human civilization changed its course to have a meteoric growth to achieve a utopia of best possible human life. It has happened and still happening by the outstanding development of science and subsequent technology. Interestingly, most of such developments are concentrated around Industrialization and urbanization. Unfortunately such developments leave aside about 70% people of this planet and shrink drastically the area of land fit for agriculture. Now, at this juncture, this development phenomenon actually raises question on future development of Human civilization. Thus, being confined in an acute transitional stage of development of human society, every man with sensible mind is worried about his future generations, consciously or sub-consciously.

Quality of life changes each day almost in every sphere. In this respect, quality food production is the most laggard subject that cannot grow with the demand and requirement of the society, compared to the subjects like entertainment, housing, health-care, transport, clothing etc.

Since last century, the wise men of our society are trying their level best to speed up the growth rate of food production mostly in respect of quantity. **Erratic application of chemicals** (leads toxicity in food), **mechanization** of cultivation practices (leads unemployment), and **genetic improvement** of crops (extinct genetic diversity) are the three major tools introduced and developed to augment quantity of production. But, quality and sustainability of crop production have not been addressed properly.

Now, the demand for development of food production in the next few decades will be concentrated on quality, nutrient supply and sustainability of the growth. Production of the quickly perishable food items will face more serious problem in this aspect of development. In this situation "greenhouse technology" or "protected cultivation technology" is the only answer that can meet up the said future demand of balance growth of production of perishable food items utilizing least possible area employing more manpower. Greenhouse technology can increase the yield manifold and at the same time can improve the quality significantly as per demand of the market. It can also be used as an immediate tool of 'organic food production', which is otherwise possible only through the change of present cultivation system, but that, will require time.

Sensing the commercial opportunity of greenhouse cultivation many far-sighted rich people around the globe has invaded this sector. They identify and copy 'greenhouse technology' with the help of experts of the pioneer countries, like Holland and Israel, without giving necessary impetus on the science behind that technology. Thus, they are dependent and have low rate of initial success. Their commercial or conservative outlook force them to cut off the spreading of these hired technology to the common mass including the educationists, extension workers, students, researchers etc. Naturally, improvement of this 'copied technology' or innovation of 'new' technology in accordance to the local situation is not possible effectively. On the other hand scientific information about greenhouse technology' in the form of text book is not available due to the same reason.

This above situation inspired me to write this book with a motto to disseminate the scientific and corresponding technical idea of greenhouse and cultivation under it. It is very much a multidisciplinary subject and indeed very difficult to co-relate amongst different discipline. I faced serious problem to generate a holistic idea about 'greenhouse technology' that are practiced presently. Poor availability of basic information and conceptual explanation, particularly for hot & humid tropical and sub-tropical areas, created a lot of problem initially. However, books on other subjects and different periodicals published for both scientific and commercial purposes helped a lot. Apart from this, some renowned and experienced scholars and experts, especially Dr. N P S Sirohi, IARI, Pusa; Dr R G Maiti, retd. Professor of Horticulture, BCKV; and Dr. K N Tiwari, IIT, Kharagur; helped me in different aspects. Actually a good period' of time and labour was invested to conceive and assemble the basic idea of 'greenhouse technology'.

This book, with a separate chapter, is also elaborated the award winning innovative "bamboo greenhouse technology. It is an eco-friendly, low-cost, scientific greenhouse structure suitable for any grower. It was innovated by me in 2012, and I dream that each and every small and marginal farmer of this planet should have such a protective structure to practice precision farming under it. I specially acknowledge the help I received from m/s Sanhit Biosolution Pvt Ltd, Bolpur, to bring this bamboo greenhouse technology into reality.

This book targets Government extension workers, individual entrepreneurs, and greenhouse-farmers to provide a clear, handy, and basic idea about greenhouse technology.

The book is also written for the teachers, researchers and students of universities and colleges who are or would be associated with the subject 'Greenhouse technology' for academic or research purposes. In the process, I have listed a number of research possibilities in Indian condition that is needed to be done in immediate future.

The book is divided into eight chapters containing 1) Introduction, 2) Basic idea, 3) Factors 4) Design, 5) Bamboo greenhouse, 6) climate control, 7) crop husbandry, and 8) few crop-wise package of practice.

Completion of this work was a difficult task. It would not have been possible without the help of my well wishers and family members. I am grateful to every one of them. I specially acknowledge the help I received continuously from Dr. M Pandit. I also acknowledge my indebtedness to all the writers of books and periodicals I have consulted during preparation of the manuscript of the present book.

I am particularly thankful to my wife Susmita, without whose devoted effort it would not have been possible to complete the manuscript. In spite of her packed daily routine she took out time to go through the entire manuscript in detail and did the necessary' alteration, correction and modification quite speedily. Her modification work frequently had forced me to rethink many issues, and thus gave me scope to improve the contents.

However, I am alert that there are a number of shortcomings in this book. At present, I am not in a position to identify and improve the same. Hope everyone who is interested about greenhouse technology will accommodate these shortcomings and also feel free to send suggestion, modification, changes, additions and alterations to the author for the improvement of this book in next edition.

Finally I convey my gratitude to 'New India Publishing Agency, New Delhi' for taking pain to publish this important book with due importance.

Dated: 11/08/2017   **Arupratan Ghosh**
West Bengal

Contents

List of Figures

List of Tables

List of Boxes

Glossary

A

Absolute Humidity: The mass of water-vapour present per unit of air (gms/m^3).

Absorption: (1) It is a surface phenomenon by which ions or compounds are taken up by the surface of solid substances/particles. (2) The net movement of water and solutes from outside of a cell or an organism to its interior.

Acid: It is the medium/substance whose aqueous solution turn blue litmus paper red (pH is less than 7) and whose hydrogen ion (H$^+$) concentration is more than hydroxyl ion (OH$^-$)

Acrylic: A glassy thermoplastic that is incorporated in rigid plastics to enhance durability.

Aeration: The process of replacement of soil-air by atmospheric air.

Aerodynamic: A style of structural covering design that reduces the air-friction to a minimum level.

Aerosol: It is a readymade sprayer/atomizer that containing necessary pesticide /chemicals usually applied directly for selective purposes. Refilling of the same is not possible.

Aggregate: A collection of soil particles gathered into a mass or cluster such as clod, crumb, or block.

Albedo: It is defined as the proportion of incoming solar radiation that is reflected. Technically it is the whiteness of or degree of reflection of incident light from an object or material.

Alkali: It is a medium whose pH is more than 7 and the hydroxyl ion (OH$^-$) concentration is more than the hydrogen ion (H$^+$).

Altitude: It is the height, measured as distance along the extended earth's radius from the mean sea level of the earth.

Ambient: Surrounding; especially, of or pertaining to the environment about a body but unaffected or undisturbed by it.

Anthesis: The bursting of pollen sac or release of pollen, otherwise the flowering period in plants.

Auxin: It is an organic compound (a class of plant growth substance or hormone) synthesized in plant body that is responsible for the growth of apical part of the plant.

Available water: It is the amount of total soil water that can be readily absorbed by plant roots. In most greenhouses (where crop is not grown on beds), at bench or pot this amount retained between suctions of 300-500cm of water.

B

Bacteria: It is a single cell microorganism that has a separate identity and do not belong to plant kingdom or animal kingdom and virtually demarcates these two.

Base: A chemical, in ionic or molecular form, capable of accepting or receiving a proton (hydrogen ion) from another substance (act as an acid).

Beam: (1) A body or support whose function is to carry and distribute lateral (perpendicular to the large dimension) loads and bending moments.
(2) A concentrated, nearly unidirectional flow of light or a propagation of electro-magnetized or acoustic wave.

Bio-control: It is the procedure of using biological agents (bacteria, fungi etc.), plants or plant-extracts to control pests (pathogens, insects, weeds).

Biological control: Use of predators, parasites, and disease causing organisms to control different pests.

Biomass: The dry weight of living matter, including stored food, or of a plant or a plant species population.

Biotic: Pertaining to life and living organisms or induced by the actions of living organisms.

Blight: Any plant disease or injury that results in general withering or decay and that leads to checking of normal growth and/or death of the plant.

Boiler: A water heater for generating steam at high pressure.

Bordeaux mixture: A fungicide made from a mixture of lime, copper sulfate and water in a certain proportion.

***Boroj*:** It is the structure (shade house made up of plant residue) under which the cultivation of betel leaves is performed. This structure is primarily used for cutting the light intensity and secondarily to maintain the necessary humidity level in favour of growth of betel-vine.

Buffer: Capacity of soil or solution to resist change of pH and/or nutritional status on application of any foreign material.

Bulk density: The weight of powdered or pulverized soil/solid material per unit of volume, generally expressed as kg/m^2., or the correct value of each

C

Calibrate: To determine the correct value of each scale reading or other device setting up for a control knob by measurement or comparison with a standard measurment.

Candle: The unit of luminous intensity, defined as $1/60^{th}$ of the luminous intensity per square centimeter of a black-body radiation at the temperature of freezing point.

Canopy: the branches, leaves and other parts of a plant or plant population that forms a spreading layer of vegetation some distance above the ground.

Compost: A well decomposed organic material produced together from different types of organic materials.

Computer: It is a device that receives, processes, and presents the two types of data e.g. analog and digital.

Conservatory: It is the structure where the plants are reared at its dormant (natural or forced) period or for hardening purpose and gearing up for next course of growth.

Cyclone: It is the in-moving high-speed air current to low pressure region of earths atmosphere, with round to elongated-oval ground plan, which certainly have upward air movement at the center, and generally have outward movement at various higher elevations in the troposphere.

D

Depression: (1) A structurally or topographically low area in the crust of earth or ground surface. (2) (Meterol.) An area of low atmospheric pressure zone, to migratory lows and troughs, also known as low, usually in a certain stage of development of tropical cyclone.

Dew point: It is the temperature in a certain atmospheric pressure at which the water-vapour of air begins to condense to water.

Diffusion: The spontaneous movement and scattering of particles (atoms and molecules) of liquids, gasses, and solids may be due to reasons like differences in concentrations etc.

Digital: Pertaining to data in the form of digit, the numeric character.

Drainage: Removal of surface-water or ground-water or excess water from soil by means of gravitational force.

Drought: A period of abnormally dry weather sufficiently prolonged so that the lack of water causes a serious sociological imbalance in the affected area in general.

Dry and Wet bulb thermometer: A humidity measurement instrument that contains two common types of thermometers, one bulb of which is covered with moist substance and the other kept normal or dry.

E

Electrical conductivity: The reciprocal of electrical resistance, which is a method for expressing salt concentration in soil or water and is expressed as *mhos*.

Electronic: Pertaining to electron devices or to circuits or systems utilizing electron devices, including electron tubes, magnetic amplifiers, transistors and other devices that do the work of electron tubes.

Emulsifiable concentrate: A material produced by dissolving/dispersing a chemical (here pesticide) and an emulsifying agent is an organic solvent.

Equator: The largest imaginary circle around the earth, equally distant from the north and South Pole that divides the earth into northern and southern hemispheres. Considering the center of earth as origin this line is pondered to be placed at $0°$.

Evaporative cooling: Cooling of any substance, particularly air, by utilizing the latent heat of vaporization of any liquid, specifically water.

Evapotranspiration: It is the total discharge of water-vapour to the atmosphere from earth's surface by evaporation and from plants by transpiration.

Exchangeable Sodium Percentage (ESP): The extent to which the exchange capacity of soil is occupied by sodium.

Exhaust fan: It is the fan used for discharge any substance, mostly gaseous and suspended materials, from any enclosed structure to outside.

F

Fertigation: It is the system of application of fertilizer through irrigation water.

Fiber-reinforced plastic (FRP): A glass fiber embedded in a plastic resin.

Field capacity: The amount of water retained in a field soil (up to considered depth) against the force of gravity after completion of drainage process.

Fogging: Fogging is the system by which **'fog'**, the ultra fine water droplets (2 to 40 micron size) suspended in air until evaporated, is generated artificially to modify a microenvironment.

Foot candle (fc): Amount of illumination produced on a surface, all points of which are at a distant of one foot from a uniform point source of one candle per square foot.

Frequency: The number of waves of electromagnetic radiation passing through a given point in an interval of unit time or number of any incidence per unit time interval.

Fungicide: Materials that repels or destroys or kill fungi.

G

Gable: The upper, triangular portion of the terminal wall of a building or greenhouse under the ridge of a sloped roof.

Galvanize: To deposit zinc on to the surface of iron or steel by the process of hot dipping or electroplating.

Genetically modified seed: The seed produced with the help of genetic engineering, the technology of production of new genes and alteration of genome by substitution or addition of new genetic material.

Graft: A piece of tissue or plant part transplanted from one individual to another or to a different place on the same individual.

Grid: A network of equally spaced lines forming squares, used for determining location of a point in a reference subject or in a map.

Gutter: A trough made up of any material and placed along the age of any structure to carry off the rain-water.

Gynoecious: Pertinent to plants that have only female or pistilate flowers.

H

Hemisphere: A half of the earth divided into north and south sections by the equator, or into an east section containing Europe, Asia, and Africa, and a west section containing the Americas.

Herbicide: A chemical used to destroy or kill unwanted plants in cropped area or weeds.

Host: The leaving plant and animal that act as a food source for pests or pests used to fed on them.

Hybrid-seed: The seed generated from genetically dissimilar parents through cross-pollination.

Hydroponics: The technique of growing plants in water or aqueous solution with all essential plant nutrients.

I

in-situ: In the original location.

Irradiance: The amount of radiant power (energy of electromagnetic radiation per unit time) per unit area that flow across or into a surface. Also known as 'radiant flux density'.

Isobars: (1) A line connecting points of equal atmospheric pressure along a given reference surface (at a specific altitude). (2) (Phys) A line connecting points of equal pressure on a graph plotting thermodynamic variable or along a given surface of a physical system.

L

Lath house: A plant protective structure, similar to shade house, made up of wood, plant residue or plastic screen (shade-net) to protect plant from excessive sunlight or frost.

Latitudes: these are the imaginary lines spread over the earth surface from east to west along the equator and characterized as the angle created at center of the earth with equator. Thus latitude of equator and poles (north & South) are 0° and 90° (North & South) respectively.

Leaching: The process of downward removal of soluble materials from soil by percolating water.

Leaf Area Index (LAI): It represents the total area of leaf per m^2 of a plant or a plant population.

Long wave radiation: It is the radiation at wavelength longer than 5000nm and also termed as thermal radiation.

Lux: Illumination on a surface of one m^2 at a distance of one meter from a uniform source of one candilla. In case of sunshine, Lux is the amount of solar radiation received (Watt/m^2) multiplied by a factor of 100.

M

Maturity: Stage of full growth or development or ripening. In case of animal kingdom it is a concept of individual but in case of plant kingdom it is often a concept of parts of an individual.

Micro-climate: It is the separated air space attached to earth surface and bounded by outline of a vegetative growth and the characteristics of that air space is different than that of the outside ambient climate.

Misting: Application of water in air with fine droplets (av. drop size 40 to 100 micron) through a specially designed device called mister.

Miticide: The chemical/pesticide that can effectively control the specific pest mite.

Monoecious: A plant having both male and female flowers.

Mulching: Covering of soil surface with any material like plant residue, plastic etc.

N

Node: A physiologically well-defined section of stem, appearing in certain intervals, from where branch, leaf and/or bud arise.

Nursery: It is the farm where plant materials are produced for onward planting into the field.

P

Pascal (Pa): A unit of pressure equal to the pressure resulting from a force of 1 (one) newton acting uniformly over an area of 1 square meter.

Pathogen: Organisms, like fungus, bacteria, virus etc., producing disease to any living material.

Perlite: A gray-white silicaceous material of volcanic origin, very light in weight, can hold water to the extent of 3-4 times of its weight and sterile in character.

Petiole: The stock that attaches the lamina or leaf blade to the stem.

pH: An expression of negative logarithm of hydrogen ion concentration, generally used to measure acidity or alkalinity of a solution. pH 7 is considered as neutral, higher and lower of this value is considered as alkaline and acidic respectively.

Photon: It is the elementary particles of light energy, which has no mass, no charge, and carries electric and magnetic forces. Actually it is the light or, more generally, electromagnetic radiation.

Photosynthesis: It is the process by which green parts (having chlorophyll) use sunlight to synthesize organic compounds from carbon dioxide and water.

Photosynthetes: The bio-chemicals that produced by the process of photosynthesis and stored in plant body for future growth.

Pollination: Transfer of pollen from the anthers to stigma of ovary of a flower.

ppm (parts per million): It represents the concentration of a liquid/solute in a water solution and expressed as number of units per million of units like milligrams per liter.

Precipitation: It is the fall of water from atmosphere in liquid or solid form into the ground surface e.g. rainfall, snowfall.

Pro-tray: This is the plastic tray having cups of specific size, called cells, grooved inside the tray to raise seedlings from an individual seed suitable for the cell.

Purlin: A horizontal part of the roof of a greenhouse structure supporting the individual structural frame of the roof.

R

Radiation: The transfer of energy in the form of electromagnetic waves of photons, which may be sunlight, ultraviolet rays and infrared etc.

Rainwater harvesting: It is the process of collection and storing of rainwater, for future use, in an area, which otherwise gets lost by run-off beyond that area.

Relative humidity: The ratio between actual vapour present in the air and the amount of vapour it could hold at saturation at the same temperature. The value is expressed as percentage.

Ridge: The highest part of an undulating structure/feature, including greenhouse roof.

S

Saline soil: Soil containing sufficient soluble salts to reduce plant growth.

Salts: Water soluble chemical compounds comprising cations (calcium, sodium, potassium, etc.) and anions (sulphates, nitrates, bicarbonates, etc.).

Saturated soil: It is the state of a soil when all the air-spaces of that soil is filled up by water.

Sensor: A device that senses either the absolute value or a change in a physical quantity such as temperature, pressure, light, sound, flow rate of gases, pH etc and convert that changes into a useful input signal for an information gathering system.

Shelter belt:

Silt: Soil particles of 0.05 to 0.002 mm diameter.

Short-wave radiation: It is generally the radiation at wavelength shorter than 5000nm and some times referred to as visible radiator (sunlight).

Sunflecks: Sunflecks are brief increases in solar irradiance that occur in under stories of an ecosystem when sunlight is able to directly reach the ground. They are caused by either wind moving branches and/or leaves in the canopy or as the **sun** moves during the day.

Sphagnum moss: It is dehydrated young residue or living parts of acid bog plants eg *Spagnum papillosum, S. capillaceum* and *S. paslustre*. This is a good root medium, which is relatively sterile, light in weight, and has very high water holding capacity.

Systemic pesticide: Chemicals that control pest attack by entering into the plant system and making the plant toxic to the pests.

T

Thermocouple: A device consisting of junction of two dissimilar elements, the electrical resistance of which is related to the temperature where they are subjected.

Tissue culture: The maintenance and growing of plant tissue (cells, callus, protoplast) and organs (stems, roots, embryos) in a nutrient medium culture or *in vitro*.

Topography: It is the high and low of land surface in respect of a reference plane normally the mean sea level.

Transitivity: The transmission capacity of a body/material on which a luminous/ light energy or radiant energy is incident upon.

Tropistic: Growing or elongation of plant cell or tissue positioned in the darker side of a plant.

Truss: Main supporting structure of a roof that is commonly supported at the ends by sidewall posts or gutters.

U

Ultra violet (UV) ray: It is an electromagnetic radiation with a wavelength from 10 nm to 400 nm, shorter than that of visible light but longer than X-rays. UV radiation constitutes about 10% of the total light output of the Sun, and is thus present in sunlight.

V

Vector: Any carrier that transfers a pathogen from one host to another.

Ventilation: The permanent open space in roof or wall of a covered structure through which the inside air (generally hot air) will pass out to outside.

Virus: Smallest and simplest living organism without any cellular structure containing a core of nucleic acid (mostly DNA) surrounded by a thin layer of protein.

W

Water use efficiency: It is the yield of crop or the crop growth achieved by the use of a given amount of applied/irrigation water.

Wilt: Loss of turgor or drooping of plant or plant parts because of inadequate water supply due to excessive transpiration or vascular plant disease.

Windbreak: A planted rows of trees or shrubs or in combination, perpendicular to the wind direction to protect crops, soil, homesteads etc. of a particular area against the adverse effect of wind.

1

Introduction

1.1. History of Cultivation

Prehistoric people lived as hunters and gatherers. About ten thousand years since now they unveiled the mystery of germination and engaged themselves in production of plant from seed. Since then it was practiced in the open field fully exposed to sun, rainfall, ambient temperature & humidity, etc. Gradually, with the development of society, they improved the package of practice and became accustomed to produce better crop in accordance with these natural conditions.

Backdrop of modern agriculture technology: Since the late 19th century, population all over the world increased rapidly, which demanded more food production. This situation prompted crop production as a profitable business. New scientific inventions were incorporated in farming practices to increase the production per unit area (yield). On the other hand, repeated cultivation of same or similar crops in the same land, year after years, gradually reduced the productivity of soil. Hence, need arose for increase of crop yield through further improvement in farming technology.

First-phase development: The incorporation of 'chemicals' into the soil as plant nutrients (fertilizers) coupled with the use of 'high-yielding varieties' changed the scenario of crop production in the sixties of last century. This, along with 'mechanization', increased crop yield many-fold and helped to bring large areas under the plough. Countries, which developed these technologies, made good amount of money/profit from this development. Gradually, the rest of the world adopted or purchased these technologies particularly to meet up the rising demand of food in their respective country.

Subsequent problem: This system of cultivation solved the food scarcity problem initially but use of chemicals, cultivation of similar crop over a large area etc. destroyed the agro- ecological balance. These technological developments, coupled with adoption of labour saving technologies, gradually created problems like gradual loss of soil health, lowering of nutrient-use-efficiency and subsequent yield. This invited severe attack of pests like insect, disease, weed and related social problem like unemployment.

As a consequence, different toxic chemicals were and are still being continuously invented to control these pests. These products, herbicide, pesticide, fungicide etc., and/or its production technology has been sold to the economically weaker countries that have no other option but to use these toxic chemicals in the field to save their high yielding crops from the attacks of different pests & diseases. As a result, environment pollution and unknown health hazard are gradually spreading all over the globe.

At the same time all these initiatives were not sufficient to fight the growing demand of, not only food, but also food supplemented with proper and balanced nutrition. However, that time it provided a temporary relief from food-insecurity or hunger, created by this development in food production system.

Second-phase of development: From the early 90's when the rate of growth of population somewhat slowed down, the nature of demand of food tilted from quantity to quality and diversity. Thus the perspective of crop selection and consequent supply of quality seed/plant material changed the farming practices significantly. Not only the quantum of demand for diverse food material increased but also the demand for quality food staff increased manifold. Developments of technology to produce quality seed/plant material in the form of hybrid-seed, genetically modified seed, tissue-cultured plant materials etc. were introduced. Apart from increasing the yield of crop these technologies were able to diversify the production range, period and quality to a great extent. To mitigate this requirement, focus was given on horticultural crops, like vegetable, fruit, flower, etc. The perishable nature of these crops and its longer harvesting period made this effort more difficult.

Backdrop of necessity of greenhouse for crop production

Increase the yield: In the mean time the pressure on agricultural land increased to accommodate urbanization, industrialization, infrastructure development etc. Naturally, to produce more in unit area in the prevailing land situation become the mandate of the production technology. But even, maximum utilization of genetic development and management procedure for open field crops, the increment in yield cannot reach beyond a certain level in the prevailing climatic situation. To overcome this restriction, the climatic factors were required to be controlled in favour of the crop, which is not possible in open field condition.

Round the year cultivation: Application of all the available technologies, mentioned above, depended on the prevalent climate thus making their use season bound. This frequently restricted the expression of optimum potential of these technologies. Most importantly these aberrant climates restricted men from cultivation of particular crops in non-specific season. On the other hand, the market demand of these products of farming was gradually created throughout the year. Therefore it was necessary to grow these crops throughout the year, irrespective of seasonal changes of the climate, which was not possible in the open field.

Quality of production: Besides the above two reasons, the demand of quality farm produce has increased sharply in recent years. Whether it is flower or food stuff, the economically sound people want mainly the quality produce even against much higher price. This sought-after quality is not possible to maintain in open field condition due to climate and its changing pattern.

The only solution: Now, primarily these problems can be solved if the desired crop is grown under protection from climatic hazards. Thus, the concept of 'covered-cultivation' or 'greenhouse-cultivation' came into existence with suitable micro-climate for a crop, which led to a new type of 'greenhouse technology'. This system to provide new cultivation infrastructure gradually gained importance and was steadily popularized.

This type of cultivation system already existed (mostly glass-house) in some agriculturally advanced countries situated in cooler regions. It was also prevalent in some other parts of the world (mostly as shade house). All these were practiced as hobby or for emergency purposes. However, its popularization and commercialization has taken place only in recent times particularly after introduction of plastic.

In India, protected cultivation existed for many centuries, particularly for the cultivation of betel leaves. This particular structure (shade house) is called 'Boroj' and is chiefly used for cutting the light intensity and secondarily to maintain the necessary humidity level, providing a suitable micro-climate.

1.2. What is Greenhouse

A 'greenhouse' is generally a house like structure where the natural sunlight, humidity, temperature etc can be manipulated to meet the requirement of growing better crop. It is basically a concept of covering the crop to create the necessary crop-microenvironment.

In cooler climate this concept somewhat resembles the 'Greenhouse effect' on earth, and from that idea this terminology has evolved. Now-a-days, when this concept is being tried in other climatic areas the term greenhouse is frequently mentioned as 'poly-house', 'protected-structure' etc.

BOX - 1

Greenhouse Effect

Mr. Joseph Fourier, in 1824 first stated that a layer of specific gases, individually called 'greenhouse gas', covers the earth. This gas mostly comprises of Carbon-dioxide (CO_2), which partially traps heat energy emitted by earth, thus maintaining the temperature in such a way that the plants on earth can grow properly. This phenomenon was termed as "greenhouse effect" and Svante Arrhenius (1896) first investigated it quantitatively, that described it as the process by which an atmosphere warms a planet.

This is warming of the atmosphere near the surface of earth caused by the absorption and trapping of long wave re-radiation (heat) emitted from the earth's surface, which obviously helps better plant growth.

The greenhouse gas consists of Carbon-dioxide (CO2), Methane, Chlorofluorocarbon (CFCs), Nitrous oxides (NO2) and others. CO2 is estimated to account for 50% of the greenhouse effect, methane for 20%, CFCs for 14% and the remainder by other components including water vapour. So the concentration of CO2 in the atmosphere is the balancing factor for steady and favourable temperature of the globe, which is balanced by the nature through its plants.

These gases act as a shield of glass pane that allow solar (short wave length) radiation, which ultimately strike the earth surface. Some of the radiation absorbed by earth thus warms the earth surface in daytime. The earth surface then emits the radiant energy (long wave-length) in to the atmosphere, and a portion of which (re-radiation) returns back to earth

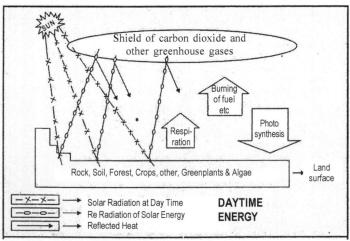

Fig. 1: Greenhouse effect

surface by the said layer of greenhouse gases at night, thus warm this planet. The re-radiation emitted by earth is less energetic and a portion of it is infrared radiation. This heat energy maintains the temperature of earth in favour of life.

Without this phenomenon, the average surface temperature of earth would be about 255°K or -18°C, thus it will be a frozen world.

On the other hand if the status of greenhouse gases increases in the atmosphere (in 2002 the CO2 % of air was recorded as 0.035 instead of 0.03) the excessive heating (global warming) of the earth would be the consequence. As a result severe drought, melting of ice resulting in rise of mean sea level (inundation) etc will occur. Besides, several other seasonal modification of climate may also take place.

Using this concept, as basic idea, a microclimate can be created artificially for best possible growth of plant in comparison to open field situation. Off season and more production (both quality and quantity) is the obvious goal for creation of said microclimate. The structure required to create such microclimate of crop is identified as greenhouse, and the technology required to generate the desired crop-microclimate is termed as greenhouse technology.

Now to elaborate this concept we may define the greenhouse, as **"a house like structure that allows a specific portion of sunlight inside and provides favorable micro crop-climate in respect of temperature, humidity, rain, and wind for optimum growth of the crop/plant grown inside"**.

However, it is not easy to create the above stated condition. It requires high precision technical knowledge of (1) construction of greenhouse, (2) climate controlling systems, and (3) best method of undercover crop husbandry. The technicality of these three aspects in a coordinated manner and application of the same can create an ideal greenhouse for growing specific plants/crops.

1.3. Brief History of Greenhouse

The idea of growing plants in environmentally controlled areas exists since Roman times. The Roman emperor *Tiberius* ate a cucumber-like vegetable daily. The Roman gardeners used artificial methods (similar to the greenhouse system) of growing to have it available for his table every day of the year. Cucumbers were planted in wheeled carts which were put in the sun daily, then taken inside to keep them warm at night. The cucumbers were stored under frames or in cucumber houses glazed with oiled cloth.

Since 17th century hobby and emergency cultivation in snow falling/winter season at cooler countries (mainly European countries) forced their men to cultivate crops under glass-greenhouses. That Glasshouse system was the basic concept of today's greenhouse/polyhouse/protected cultivation.

The French botanist Charles Lucien Bonaparte is often credited with building the first practical modern greenhouse in Leiden, Holland, during the 1800s to grow medicinal tropical plants

Gradually that concept or technology was being used in different forms for different situation and purposes to adopt larger commercial greenhouse since end of the 19th century. Netherland or Holland is the first country to take the lead to implement large scale commercial greenhouse projects, mostly for flower production.

Mainly due to very high initial cost, this 'glass-greenhouse' era did not last beyond the middle of 20th century. Introduction of flexible polyethylene film and other plastic materials, as cheaper covering material, gradually replaced the glass.

In 1930s the polyethylene film was developed in England. Around the middle of 1950s the plastic greenhouses (Polyhouse) came into practice, and again Holland took the lead to adopt it on a commercial scale. For 25 years the area covered under plastic greenhouse has increased significantly and steadily. Only in USA the area of plastic greenhouse cultivation grew from 40 to 2300 to 4800 to 9100 acres in the mid-50s to mid-60s to 1977 to 2011 respectively.

At the end of 20th century Israel made revolutionary growth with their greenhouse technology suitable for Hot and dry climate. They grew from 100 acres to 13000 acres within 15 years (1985 to 1999).

Gradually the greenhouse technology is spreading everywhere. Not only the developed countries and the countries of cooler climates, it is spreading almost in each developing country, particularly who are agriculturally active irrespective of climatic condition.

In the world, in the present status of greenhouses, China takes first place, South Korea, Spain and Japan follow respectively and Turkey occupies fifth place. Countries like Morocco shows a good growth in area covered under greenhouse. On the other hand initial adopters like Italy, France, UK, etc are in somewhat in a stagnant situation in respect of its growth.

Problems of spreading of greenhouse: Use of plastic greenhouse in crop production has shown phenomenal development due to its wide adaptability in wide range of climatic conditions and efficient ability to control climate. Apart from the demand for quality produce, the high market price of off-season farm produce, especially flowers, is the reason for which this present greenhouse/polyhouse cultivation system is spreading fast.

However, the main problem to adopt this system of cultivation by the common and small farmers of economically weaker countries is its initial expenditure and tricky technical specifications. The technical specifications were solely based on the local climatic situation for which it is not possible to copy the technology of other countries with different climatic situation and implement it elsewhere.

1.4. Scenario of Greenhouse Cultivation

Status of greenhouse around the globe: Greenhouse technologies are being commercially utilized all over the world since last few decades. However, the level and extent of their use is different in different countries. Thus, it is necessary to discuss the status of greenhouse cultivation around the world and its future prospect. These structures may be of different types according to the prevailing climate and its use. It includes high-technology (fully and automatically controlled), medium- technology (partially and manually controlled), low-technology (naturally controlled), and Plastic tunnel greenhouses.

There has been tremendous increase in area under greenhouse cultivation in most of the countries (Table-1). As history indicates, in cool climates greenhouse cultivation has spread more and despite having a lot of potential, the medium and low technology greenhouses has not spread properly. However, the areas under plastic tunnels, including walk-in-tunnels, are increasing with great pace. Countries like China, Japan, Spain, Italy, Turkey, Morocco are showing very good growth.

In recent times, more than 55 countries in the world are engaged in commercial cultivation under cover/greenhouse in a large scale and it is growing fast.

Status of recent past

It is very difficult to collect true picture of the area of greenhouse of different countries. The respective govt. data not always show the actual figure due to the improper collection system of information regarding new and abandoned greenhouses of farmers. However data related to State sponsored greenhouses can give us an idea about it.

Netherland, the oldest country using greenhouse for commercial cultivation, had greenhouse area of about **89,600 ha**, in the most advanced form. It equally shares its greenhouse area for flower and vegetable cultivation. They earn US$ 45000/year from 1000m^2 of greenhouse by cultivating flowers.

In **Spain 28,000 ha** has been covered under greenhouse of which most are naturally ventilated type and are for growing vegetables like tomato, cucumber, capsicum, etc. Earlier they used Low-plastic tunnels for growing strawberries and melons, which reduced significantly in recent past.

Italy uses its **19,500 ha** greenhouse structures mostly for vegetable cultivation generally for own consumption. Only half of its greenhouses have heating arrangements.

The **USA** has a total area of about **15,000** ha under greenhouse mostly used for flower cultivation with a turnover of 3.4 bilion USD per annum and the area is growing considerably.

Israel uses most sophisticated greenhouse technology, which give them the third rank in cut-flower export in 1987 with a limited greenhouse area. In recent past they utilized about **18,000 ha** sophisticated greenhouses for production and export of flower, vegetables, and fruits to Europe in off season. They use evaporative cooling system with precision to control the climate of greenhouse.

Turkey (12,000 ha), **Morocco, Kenya,** and **Algeria** are expanding their greenhouse area rapidly.

In Asia, China, Japan and recently South Korea are the leading users of greenhouse technology of the world. They uses greenhouse primarily to extend the growing season, both in spring and fall.

The development and spread of greenhouse technology in **China** has been faster than any other country in the world. The area under greenhouse has already been increased to **51,000 ha** and 1/5th of this is used for fruits like grapes, cherry, fig, loquat, lemon, mango, etc. They used every type of greenhouse technology, from low-cost to hi-tech, to construct greenhouses. It is increasing at the fastest rate in the world and would cross **100,000 ha** shortly.

Japan has about **40,000 ha** greenhouse area, of which 22% is for cultivation of fruit orchards. More than sixty percent of greenhouse is utilized for wide range of vegetable and flower cultivation. Most of their own vegetable consumption is being met from these greenhouses.

Country like **South Korea** has about **21,000 ha** under greenhouse for the production of fruits and flower.

Scenario of Greenhouse Cultivation in India

Commercially greenhouse technology in India stepped in its infancy in 1990s. During those periods several large naturally ventilated greenhouse units came up, copying the foreign technology and collaborating with foreign agencies, to produce flowers for export. The total area under greenhouse at that point of time was only **100 ha**.

In recent past (2000 to 2010) large, medium and small, gutter connected and stand-alone, naturally ventilated greenhouse and walk-in-plastic tunnels of smaller size has been created in many parts of India. In 2012 the area increased to 5730 ha only.

The scenario was changed afterwards gradually, particularly after govt. was taking the initiative through different subsidy schemes. Now it is growing throughout India, but at a slow pace due to climatic situation and high initial investment. Even, Leh of Jammu & Kashmir alone boasts of more than 20000 small units (50 m² each, i.e., 100ha in total) of greenhouses. Mostly fresh vegetables are grown in these small greenhouses and these low-technology structures have permitted the extension of growing season by four to six months.

As on today, with due support of Government this technology is spreading throughout the Country for cultivation of quality flowers and vegetables and also for nurseries. Production of seedling is the other field where the greenhouse technology can be utilized in India.

Leading Vegetable Producing States in Terms of Production (2013-14)
India's contribution to world flower basket was a meagre 0.3%. In case of cut-flower only it is further low. In respect of greenhouse vegetable this status is literally zero. However, there is a huge scope for strengthening its position in the international flower market due to its range of climatic situations and developing greenhouse technology suitable to these climates.

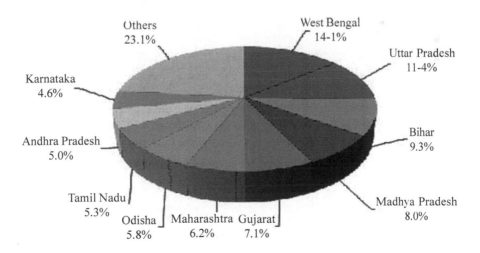

Major Cut Flower Producing State's 2010-11

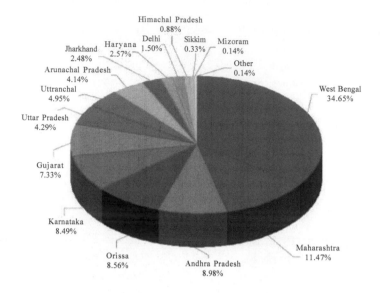

Area and Production of Flowers in India

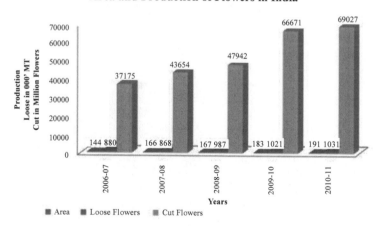

Source: Indian Horticulture Database-2011

Fig: 2: Status of Indian states in area of vegetable and flower production

Over and above this, Quality Gerbera, orchids and chrysanthemum, cultivated here under greenhouse condition, have large demands in international markets. To exploit this potential to its full extent, larger number of outbound international flights, infrastructure like organised flower markets and cold chain facility is needed.

Table 1: Export of flower (including collected dry-flowers/leafs etc) from India

Country	2008-2009		2009-2010		2010-2011	
	Qty	Value	Qty	Value	Qty	Value
United States	7.111.5	7.213.5	5871.1	5.305.6	7153.9	5686.5
Germany	3.589.8	3.966.3	36X8.2	4065.0	4511.6	4280.6
Netherland	4.640.2	5.987.2	3146.3	42179	2989.5	41620
United Kingdom	4.369.7	4.284.5	3707.3	3788.3	4116.0	3761.8
Japan	965.3	1.791.0	970.9	1558.7	576.7	1.151.9
United Arab Emirates	762.7	991.8	971.7	1071.1	812.8	959.1
Italy	1268.3	1.373.3	1453.7	714.4	1.234.0	856.8
Canada	782.4	1.135.3	534.1	769 0	532.8	798.4
Belgium	1.084.3	845.5	470.2	484.4	762.0	774.3
Ethiopia	99.0	1.657.0	706.2	1.746.7	69.1	633.5
Other	6.125.4	7.636.0	5294.4	5.625.6	5.017.8	5.580.5
Total	30.789.3	36.881.4	26.814.5	29.446.4	27,776.2	28.645.4

Still it is an early stage of development of greenhouse technology in India and it will gain its momentum only when small farmers will take up this of greenhouse cultivation technology into their stride.

2

General Idea About Greenhouse Technology

2.1. What is Greenhouse Technology?

Greenhouse technology involves primarily the structural design that can take care of climate control mechanisms to create proper micro-crop-climate and secondarily it involves the technology needed for optimum growth and production of crop grown under this structure. Combination of these two aspects may be expressed as 'greenhouse technology'.

So the 'greenhouse technology' may be defined as "The technology required to built a structure, covered fully, partly or porously for manipulating the inside climatic conditions, under which crops/plants can be grown with special measures".

2.1.1. Categories or characters of greenhouse

i. Depending upon the nature of covering materials, greenhouses may be divided into two broad group viz. (1) glass/plastic house and (2) net/shade house.

ii. Depending upon the cost as well as climatic control system, greenhouse may be classified into three categories viz. (1) Fully and automatically controlled or high technology greenhouse, (2) Manually controlled or medium technology greenhouse and (3) Naturally ventilated (controlled) and Low cost shelter or tunnel type greenhouse.

iii. Depending upon the purpose and type of crop/plant grown, the greenhouses may be sub-divided as (1) Hobby or experimental greenhouse, (2) Commercial cultivation greenhouse and (3) Nursery greenhouse.

Now, after all such classifications a greenhouse may be established or selected by assembling all three criteria mentioned above depending upon the necessity of viability of the project. Thus there may be many permutations & combinations available for consideration.

For example, any greenhouse should have minimum three characters taken from each of the above- mentioned group of classification, as given below:

1. Case 1 - Poly + -High-cost/technology + Commercial;
2. Case 2 - Poly + -Low-cost/technology + Nursery;
3. Case 3 - Shade/net + Medium-cost/technology + Commercial;

Nowadays, need and subsequent trials has yielded many more types of greenhouses combining the sub-groups of a basic character like covering material, control system etc. For example (1) plastic film and shade-net are often simultaneously used in greenhouses; (2) some medium technology greenhouse has the facility to be used as shelter type greenhouse in a particular season.

Both Hobby or experimental and nursery type greenhouses are multifunctional and expensive. While Hobby or experimental type is non-commercial, the greenhouse for nursery has separate character and a commercial outlook.

2.2. Basic Conceptual Planning for A Greenhouse as a Project

Decision: To start a greenhouse a well laid out basic plan is essential. Certain decisions are to be taken before venturing for such a plan. These decisions shall be linked with the following aspects.

1. Crop or Crops to be grown - seasonal flower/vegetables, nursery, raising of ornamental plant, perennial flower/vegetable/fruit, any other crop etc.
2. Growing period – It should be year-round or seasonal. If seasonal, consider the off-season utilization.
3. Where to grow plants - pots or beds or benches etc.
4. Root medium - soil-based, soil-less, hydroponics, soils.
5. Production target - Floor area of the greenhouse and number of plants.
6. Profitability of project and source of fund/capital - Proper financial appraisal and arrangement of capital from own fund and from bank etc.
7. Post harvest, Marketing and its procedure - Local or outside market and marketed by self or through other agencies.

Information: On the basis of the decision taken, the plan of the greenhouse project should be prepared. For this, the detailed information regarding the following aspects has to be collected.

1. Detail of the site of the greenhouse.
2. Detail of crop and source of plant material.
3. Estimate and design of greenhouse to be constructed.
4. Estimate of the cost of climatic control mechanism and irrigation.

5. Designing of bed or pot or benches and preparation of root media.
6. Source of inputs including human resource and its cost.
7. Registration, if required.
8. Post harvest and Marketing facility and arrangements.

Project planning: After collecting the above mentioned information a project calendar is required to be prepared to smoothen up the entire operation of the project. A model project calendar is given below for necessary consideration.

Table 2: Implementation calendar of a greenhouse project (Project Calender)

Practiculars	April				May				June				July				August		
	1 wk	2 wk	3 wk	4 wk	1 wk	2 wk	3 wk	4 wk	1 wk	2 wk	3 wk	4 wk	1 wk	2 wk	3 wk	4 wk	1 wk	2 wk	3 wk
Land Development	█																		
Greenhouse foundation		█																	
Bedding material Application: Red soil		█																	
FYM			█																
Rice Husk																			
Greenhouse Erection					█	█													
Polythene fitting									█										
Bedding Material Mixing									█										
Fumigation											█								
Bed Preparation												█							
Drip & Misting System														█					
Spraying System														█					
Hosing Tap														█					
Mixing of Neem cake Silica dust sterameal vermicompost																	█		
Mixing Besal Dose																		█	
Plantation																			█
Labour quarter									█	█	█	█	█	█	█	█			
Commencement of Flower																			

2.3. Subjects Involved in Greenhouse Technology

Greenhouse technology is a multidisciplinary subject that composes of three basic subjects namely (1) Climatology, (2) Structural engineering and |(3) Crop husbandry. It is essential to understand the role of these three subjects for planning a greenhouse, which has been discussed elaborately in the following chapters.

A brief outline of these three subjects is given below, which will help to understand the basics of greenhouse technology.

2.3.1. Climatology

Greenhouse is primarily meant for controlling or modification of a few natural climatic elements e.g. (i) sunshine, (ii) precipitation, (iii) temperature, (iv) humidity and (v) wind velocity of the area. So collection of detailed information about these climatic elements, prevailing in the locality where the greenhouse will be constructed, is the first job to select the greenhouse technology.

In general the latitude of an area can provide general information about the solar radiation received by the earth surface, which is directly or indirectly related with the status of temperature, precipitation, humidity and wind velocity of an area.

The biological activity of most of the plants is better between mean minimum and mean maximum temperature of 12°C to 35°C. So, one should try to modify or control the temperature of cropped area under greenhouse anything between 15°C to 30°C throughout the year. Along with this, a control over precipitation is essential to have a good crop particularly in the wet or humid region.

Round the year crop production in the open field is not possible primarily because of seasonal variation of different climatic elements, particularly temperature, radiation and precipitation. Thus history of seasonal variation of different climatic components must be meticulously observed and noted.

Climatologists have divided earth into several climatic zones depending on some specific climatic factors like precipitation and temperature.

But for basic greenhouse technology three predominant climatic factors rule the general character of different climate for greenhouse of this globe. The identified dominant climates considered for greenhouse technology are: (1) Cool climate, (2) Hot climate & (3) Humid climate. All these three basic type of climatic characters require completely different ideas and types of technology for greenhouse.

However, apart from these basic climates of greenhouse technology, different criteria of different climatic components and the seasonal variation as well as phenomenal seasonal changes has to be considered. Accordingly the basic greenhouse technology of that area may be modified to cope with the said variations and severity of these climatic factors.

As a result, sometime mixture of two basic types of technology is necessary. This concept will create separate technologies for Cool & Humid, Humid &

Hot, Hot & Dry climates, apart from the basic three types of greenhouse technology mentioned earlier.

2.3.2. Structural engineering

This is the subject with the help of which the structural design of a greenhouse is prepared.

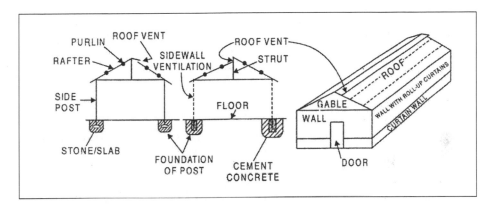

Fig. 3. Design and Structure of stand alone 'A' frame type greenhouse

It has four different items. Brief outlines of these four items are narrated below:

(i) **Super Structure**: The strength, size, shape of structure as well as the arrangement of ventilation and other climate control systems of a greenhouse is considered on the basis of the analysis of the different climatic and crop factors. Strength of super structure is calculated on the basis of 'load' analysis, where 'wind-load' is the primary factor and the 'dead-load', 'live load' and 'crop-load' are the secondary factors. (Detail is given in Chapter 4.4)

(ii) **Covering/cladding material**: The selection of covering material (from different types of material) and mechanism and device of fixing it is the next important part of the structural engineering. It may be film or net and the selection is again dependent on the climate and crop. (Detail is given in Chapter 4.3)

(iii) **The climate control mechanism**: The climate control mechanism along with planning of operation of ventilation is the third aspect of structural engineering for designing a greenhouse. This is an integral part of structural engineering and shall be considered simultaneously with designing of super structure. This includes type of ventilation, type of cooling mechanism, type of heating mechanism, type of air circulation mechanism etc. These arrangements are essential either singly or in combination for every greenhouse (details of these items are given in chapter - 6).

(iv) Protection and maintenance: Protection and proper maintenance of the structure, covering materials and machinery is the final aspect of structural engineering (details of these items are given in chapter – 4.5).

2.3.3. Crop husbandry

This subject shall be the targeted or focused area of greenhouse technology to make it financially viable. All these designing and planning of a greenhouse structure involved to select a suitable crop, which should be cultivated under the structure suitably and profitably. Thus, it is the final aspect of greenhouse technology on which its success lies. This is also the most changeable part of a greenhouse technology.

The subject of crop husbandry is dependent on the structure of greenhouse, which is designed on the basis of selected crop and climatic factors of the area. So the climate- structure- crop relationship is the sole yardstick to measure the technical quality of a greenhouse. More coordinated relationship is likely to give more success. This is the reason why single-crop greenhouse is predominant and commercially successful than that of multi-crop greenhouse.

However, this subject deals with the following aspects:

1. Selection of crop and corresponding varieties,
2. Preparation of necessary root media,
3. Preparation of bed or pot or trough for growing crops,
4. Training, pruning, and canopy management,
5. Irrigation or fertigation system and its management,
6. Nutrient management,
7. Plant protection,
8. Harvesting and 1st tier post-harvest management,

Though it looks similar to the different aspects of open field cultivation, almost all these items are completely different in principal and practice under greenhouse condition.

Out of these eight aspects selection of crop and variety is the most important aspect that determine the effectiveness of greenhouse technology (detail is given in chapter - 6).

2.4. Fundamental Points of Greenhouse Technology for Different Climatic Situations

As discussed earlier, there are three basic types of climate to consider primarily for greenhouse technology, which are 1) Cool climate, 2) Hot & dry climate, and

3) Humid & hot climate. However, use of different kind of nets or porous covering materials (shade-net, Insect-proof net etc.) for greenhouse construction in hot and humid climate is also considered here as 4th segment of greenhouse technology.

Greenhouse technology for humid climate is relatively new and complicated. The necessity for isolation of humid climate is mainly because of the great effect of humidity under greenhouse condition on temperature, evaporation, transpiration, disease infestation and ultimately the crop health.

Increase of humidity reduces the re-radiation of solar energy, thus restrict reduction of night temperature. This is similar to 'Heat Index' that relates to our feeling temperature under shade.

A similar kind of relation between wind and feeling temperature also exist and has a pronounced effect in cooler climate, which is called 'wind-chill-factor' of temperature or 'Wind-chill'.

Wind chill temperature is always lower than the air temperature for values where the wind-chill formula is valid. When the apparent or feeling temperature is higher than the air temperature, the formula of heat index is used.

BOX-2

Heat Index (HI)

The **heat index (HI)** or **humiture** or **humidex** is an index that combines air temperature and relative humidity under shade, as an attempt to determine the human- perceived equivalent temperature, as how hot it would feel if the humidity were some other value. The result is also known as the "felt air temperature" or "apparent temperature". For example, when the temperature is 32 °C (90 °F) with 70% relative humidity, the heat index value is 41 °C (106 °F).

It is not uncommon for temperature to remain above 80°F (27⁰C) during summer nights if humidity is high (dew point above 65°). So, although plants do not "feel" a high heat index, they are affected by the slow temperature decline during nights of high humidity through increased respiration. Such respiration is wasteful because it stimulates respiration more than photosynthesis and burns away or oxidizes sugars that could have been stored in seeds/plant parts as yield.

Table of values of heat index: The table below is from the U.S. National Oceanic and Atmospheric Administration. The columns begin at 80 °F (27 °C), but there is also a heat index effect at 26°C and similar temperatures when there is high humidity.

Temperature (RH %)	80 °F (27 °C)	82 °F (28 °C)	84 °F (29 °C)	86 °F (30 °C)	88 °F (31 °C)	90 °F (32 °C)	92 °F (33 °C)	94 °F (34 °C)	96 °F (36 °C)	98 °F (37 °C)	100 °F (38 °C)	102 °F (39 °C)	104 °F (40 °C)	106 °F (41 °C)	108 °F (42 °C)	110 °F (43 °C)
40	80 °F (27 °C)	81 °F (27 °C)	83 °F (28 °C)	85 °F (29 °C)	88 °F (31 °C)	91 °F (33 °C)	94 °F (34 °C)	97 °F (36 °C)	101 °F (38 °C)	105 °F (41 °C)	109 °F (43 °C)	114 °F (46 °C)	119 °F (48 °C)	124 °F (51 °C)	130 °F (54 °C)	136 °F (58 °C)
45	80 °F (27 °C)	82 °F (28 °C)	84 °F (29 °C)	87 °F (31 °C)	89 °F (32 °C)	93 °F (34 °C)	96 °F (36 °C)	100 °F (38 °C)	104 °F (40 °C)	109 °F (43 °C)	114 °F (46 °C)	119 °F (48 °C)	124 °F (51 °C)	130 °F (54 °C)	137 °F (58 °C)	
50	81 °F (27 °C)	83 °F (28 °C)	85 °F (29 °C)	88 °F (31 °C)	91 °F (33 °C)	95 °F (35 °C)	99 °F (37 °C)	103 °F (39 °C)	108 °F (42 °C)	113 °F (45 °C)	118 °F (48 °C)	124 °F (51 °C)	131 °F (55 °C)	137 °F (58 °C)		
55	81 °F (27 °C)	84 °F (29 °C)	86 °F (30 °C)	89 °F (32 °C)	93 °F (34 °C)	97 °F (36 °C)	101 °F (38 °C)	106 °F (41 °C)	112 °F (44 °C)	117 °F (47 °C)	124 °F (51 °C)	130 °F (54 °C)	137 °F (58 °C)			
60	82 °F (28 °C)	84 °F (29 °C)	88 °F (31 °C)	91 °F (33 °C)	95 °F (35 °C)	100 °F (38 °C)	105 °F (41 °C)	110 °F (43 °C)	116 °F (47 °C)	123 °F (51 °C)	129 °F (54 °C)	137 °F (58 °C)				
65	82 °F (28 °C)	85 °F (29 °C)	89 °F (32 °C)	93 °F (34 °C)	98 °F (37 °C)	103 °F (39 °C)	108 °F (42 °C)	114 °F (46 °C)	121 °F (49 °C)	128 °F (53 °C)	136 °F (58 °C)					
70	83 °F (28 °C)	86 °F (30 °C)	90 °F (32 °C)	95 °F (35 °C)	100 °F (38 °C)	105 °F (41 °C)	112 °F (44 °C)	119 °F (48 °C)	126 °F (52 °C)	134 °F (57 °C)						
75	84 °F (29 °C)	88 °F (31 °C)	92 °F (33 °C)	97 °F (36 °C)	103 °F (39 °C)	109 °F (43 °C)	116 °F (47 °C)	124 °F (51 °C)	132 °F (56 °C)							
80	84 °F (29 °C)	89 °F (32 °C)	94 °F (34 °C)	100 °F (38 °C)	106 °F (41 °C)	113 °F (45 °C)	121 °F (49 °C)	129 °F (54 °C)								
85	85 °F (29 °C)	90 °F (32 °C)	96 °F (36 °C)	102 °F (39 °C)	110 °F (43 °C)	117 °F (47 °C)	126 °F (52 °C)	135 °F (57 °C)								
90	86 °F (30 °C)	91 °F (33 °C)	98 °F (37 °C)	105 °F (41 °C)	113 °F (45 °C)	122 °F (50 °C)	131 °F (55 °C)									
95	86 °F (30 °C)	93 °F (34 °C)	100 °F (38 °C)	108 °F (42 °C)	117 °F (47 °C)	127 °F (53 °C)										
100	87 °F (31 °C)	95 °F (35 °C)	103 °F (39 °C)	112 °F (44 °C)	121 °F (49 °C)	132 °F (56 °C)										

BOX- 3
Wind Chill Factor

Wind-chill (popularly **wind chill factor**) is the perceived decrease in air temperature felt by the body on exposed skin due to the flow of air. The process (heat transfer) convection from a warm surface heats the air around it, and an insulating boundary layer of warm air forms against that surface, thus restrict further heat loss. Moving air disrupts this boundary layer, or epiclimate, allowing for relatively cooler air to replace the warm air against the surface. The faster the wind speed, the more readily the surface cools.

The effect of wind chill is to increase the rate of heat loss and reduce any warmer objects to the ambient temperature more quickly. Dry air cannot, however, reduce the temperature of these objects below the ambient temperature, no matter how great the wind velocity. For most biological organisms, the physiological response is to generate more heat in order to maintain a surface temperature in an acceptable range. The attempt to maintain a given surface temperature in an environment of faster heat loss results in both the perception of lower temperatures and an actual greater heat loss.

Wind-chill temperature is generally considered only for temperatures at or below 10 °C (50 °F) and wind speeds above 4.8 kilometres per hour (3.0 mph). As the air temperature falls, the chilling effect of any wind that is present increases. For example, a 16 km/h (9.9 mph) wind will lower the apparent temperature by a wider margin at an air temperature of "20 °C ("4 °F), than a wind of the same speed would if the air temperature were "10 °C (14 °F).

Temperature (°F)

Calm	40	35	30	25	20	15	10	5	0	-5	-10	-15	-20	-25	-30	-35	-40	-45
5	36	31	25	19	13	7	1	-5	-11	-16	-22	-28	-34	-40	-46	-52	-57	-63
10	34	27	21	15	9	3	-4	-10	-16	-22	-28	-35	-41	-47	-53	-59	-66	-72
15	32	25	19	13	6	0	-7	-13	-19	-26	-32	-39	-45	-51	-58	-64	-71	-77
20	30	24	17	11	4	-2	-9	-15	-22	-29	-35	-42	-48	-55	-61	-68	-74	-81
25	29	23	16	9	3	-4	-11	-17	-24	-31	-37	-44	-51	-58	-64	-71	-78	-84
30	28	22	15	8	1	-5	-12	-19	-26	-33	-39	-46	-53	-60	-67	-73	-80	-87
35	28	21	14	7	0	-7	-14	-21	-27	-34	-41	-48	-55	-62	-69	-76	-82	-89
40	27	20	13	6	-1	-8	-15	-22	-29	-36	-43	-50	-57	-64	-71	-78	-84	-91
45	26	19	12	5	-2	-9	-16	-23	-30	-37	-44	-51	-58	-65	-72	-79	-86	-93
50	26	19	12	4	-3	-10	-17	-24	-31	-38	-45	-52	-60	-67	-74	-81	-88	-95
55	25	18	11	4	-3	-11	-18	-25	-32	-39	-46	-54	-61	-68	-75	-82	-89	-97
60	25	17	10	3	-4	-11	-19	-26	-33	-40	-48	-55	-62	-69	-76	-84	-91	-98

Wind (mph)

Frostbite Times   30 minutes   10 minutes   5 minutes

Wind Chill (°F) = 35.74 + 0.6215T - 35.75(V^{0.16}) + 0.4275T(V^{0.16})

Where, T= Air Temperature (°F) V= Wind Speed (mph)   Effective 11/01/01

It is very difficult to isolate wet/humid climatic condition from cool or hot climate. Here very high and prolonged precipitation along with high temperature is the basic source of high humidity, which actually increase 'heat index'. The tropical and sub-tropical areas having lower difference between day & night temperature and prolonged rainfall (for about 5 moths) with plentiful surface and soil water created the humid climate, which provides a very high humidity in the air.

Thus, we have to consider some basic points for adoption of greenhouse technology necessary for these specific climatic conditions.

2.4.1. Basic points to be followed in cool climate area

1. It should be completely closed to catch hold the daytime solar radiation inside the house to increase the inside temperature at day time and maintain the higher inside temperature at night.
2. The structural strength shall be more and the roof shall be designed properly to glide and withstand the probable load of snow.
3. The height of the greenhouse may be low, with closable ventilation at the top. These greenhouses will be gothic for round/tunnel type or saddle-roof type with necessary slope to slip snow.
4. Permanent heating arrangement is must for such greenhouse.
5. The covering material should be transparent, thick, and rigid or double layered, in case of poly-film, with a low capacity of energy transmission to reduce the heat loss from inside.
6. Good inside and outside drainage system of such greenhouse should be considered in proper manner.

2.4.2. Basic points to be followed in hot & dry climate area

1. The height of greenhouse shall be more with larger ventilator at the top, which increases the chimney effect of the greenhouse to vent out the inside hot air. This will also increase the ratio of volume to surface area of the greenhouse, which reduces the inside heating effect of solar radiation.
2. The greenhouse should have sufficient side-wall ventilation. The percentage of total ventilation area to floor area may be as high as 60% starting from 20%. If required, closeable or roll-up curtain may be provided in the ventilation to manage low outside night temperature during winter.
3. Thus, it may be a 'naturally ventilated greenhouse' which can control the rate of heat exchange between inside and outside climates through these closable ventilation systems, with different day and night temperature.

4. A separate and permanent evaporative cooling arrangement like 'fan-pad cooling' or 'fogging' system should be provided to all such greenhouses to reduce excessive heat and simultaneously increase inside humidity.

5. The air-circulating system may also be installed along with fogging system for better plant growth.

6. Though the roof load is not much, the structural design should have the strength to bear the high wind load @ 10 kg/m². Thus, it can withstand wind pressure of about 120 km/hr.

7. Windbreak is a must for such greenhouse to create barrier against hot wind desiccation.

8. Proper budgeting of water is to be done and water use efficiency should be ensured in the greenhouse.

9. Application of thermal-screen or shade-net in the greenhouse to reduce the entry of excess solar radiation should be made. 50% shade net can reduce 20% of temperature during hottest period at daytime. This will not hamper the growth of crop; instead it will remove the adverse effect of temporary wilting of the crop.

10. Rainwater harvesting structure will be created to harvest rainwater from roof of the greenhouse and will be reused for irrigation. It will also take care of the drainage.

2.4.3. Basic points to be followed in wet/humid & hot climate

1. Here predominantly the 'naturally ventilated' greenhouse is used to control high humidity and temperature. The height of greenhouse should be more and the larger roof vent is must to create proper chimney effect.

2. Most of the sidewall should be open and may be covered with fixed or removable insect-proof net. The total ventilation should be more than 60% of the floor area and sometimes it may be more than floor area to induce natural ventilation.

3. Insect proof nets may be used to protect the openings of the greenhouse from insects particularly the vectors of different viral diseases, which on other hand may reduce the natural ventilation and as a result increase inside temperature.

4. The airflow inside the greenhouse shall be increased through air-circulating fans. When outside humidity is very high the exhaust fan with tube system (fan-tube ventilation) is good to increase air-flow and to reduce the inside humidity in favour of crop growth.

5. Shelter structure or greenhouse with permanently open side-wall use natural airflow to reduce canopy level humidity and temperature. This

principle may be used for greenhouses in this type of tropical climate, where difference between day and night temperatures is less.

6. Reduction of evaporation from soil surface inside the greenhouse shall be done through mulching along with drip irrigation.

7. Very good outside drainage system of the greenhouse shall be provided carefully, otherwise seepage will create a serious problem.

8. In dry hot summer, evaporative cooling method may be used through misting or fogging, whenever necessary, to cool down the high day time temperature.

2.4.4. Basic points to be followed for shade/net-house in hot & humid climate

1. Leaves/twigs/plant materials or plastic shade-net and IP-net is used as covering material for such greenhouses. (It is the oldest type of cover cultivation practiced by us for betel-leaf cultivation etc. Now we are using such structures for different purposes.)

2. It shall be permeable to rain water but, at the same time, shall reduce the heavy impact of rain on soil & crop.

3. Horizontal pattern of roof does not require strong roof structure. Wire rope may be used to hold the covering material at the top. The shade/net house along with pole or posts is designed in such a way that it can bear the impact of wind and rain.

4. 25 to 75% shading net or 40 to 70 mesh insect proof net (IP-net) as per necessity is generally used as covering material for such structure.

5. The disease and pest incidence is more in humid area. To protect the crop a **net-house** with **IP-net** with good drainage facility, similar to that of shade-house, is very much useful.

Note

All the above points are general in nature considering only the basic plastic greenhouse/net-house, without considering local climate & other situation, crop, socio-economic condition, raw material availability, and finally the financial viability. The structural material used in the greenhouse and shape of greenhouse is also not specified in these points.

2.5. Purpose Oriented Greenhouse Technology

As discussed in the introduction, there are different ways to classify the model of greenhouse. The experimentation for designing greenhouses for new locations evolved new models of greenhouses depending upon the local factors as described earlier.

Depending upon the basic purposes, the greenhouse can be primarily divided into two broad groups, (1) Greenhouse for hobby or nursery or research purpose and (2) Greenhouse for cultivation purpose. Here, in this book, we focus on the greenhouses for cultivation purposes. However, brief outlines of both these types are narrated as follows:

2.5.1. Greenhouse for hobby or urban farming, nursery and research

In glass-greenhouse era hobby greenhouses existed predominantly in cool European countries. After that, in polyhouse era, small hobby greenhouses covered with plastic film or rigid-plastic were created and used throughout the world, particularly in the urban areas.

Actually, the demand of these urban greenhouses is growing profusely. The sizes of such urban greenhouses are commonly 20 to 100 sqm. These urban greenhouses is sometimes termed as 'kitchen garden greenhouse', and are gradually become popular among the unban people. In such cases, and also for hobby/urban farming, the technology of 'Vertical Farming' is being used frequently, particularly to accommodate higher number of plant without compromising the growth.

Thereafter, in other areas besides cool European countries, the small single-span greenhouses made of plastic were mostly used for raising seedling, cutting, hardening of delicate plant materials, rearing of ornamental plants (may be under shade-house or poly-net greenhouse) etc. All these activities are part and parcel of Nursery and different kind of greenhouse structures are essential for that purpose.

Apart from these hobby, urban, and nursery greenhouses, there are experimental or research purpose greenhouse, which is covered with glass or polycarbonate or plastic film. These greenhouses were used frequently by different institutes and organizations spread over several countries of this globe.

These types of greenhouses are generally smaller in size, may be of 20 to 200 m², and may be of single-span structure. In cool climate the small size closed greenhouse works better, where the surface area to volume ratio is low. This increases the inside temperature rapidly that is suitable for cooler areas. But in warmer areas such small size closed structure is not suitable, as it frequently increases the inside temperature beyond biological limit. So in such areas the greenhouses shall be taller and provided with necessary open surface area for ventilation.

In commercial nursery, the specially designed poly greenhouse / shade house are used for rearing or hardening of plantlets or plants (sometime such structures are called as conservatory). The size of such greenhouse shall be bigger and may be multi-span structure. The A-shaped even or broken span roof, (Fig. 3) with or without vent at the top, is generally used for these greenhouses. These constructions are to be made proportionately so that the space is well utilized for convenient walkways and for beds/branches. These structures vary from elementary to home-constructed to most modem installations. In comparison, these structures are more strong and stable to last long, and for that, sometime polycarbonate sheets are used as covering material instead of Polyethylene film.

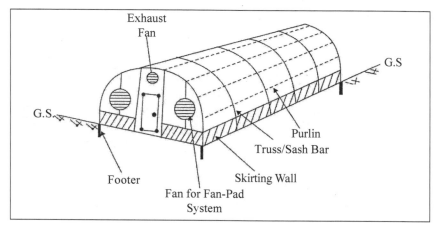

Fig. 4: Quonset type greenhouse for nursery & experiment

Nowadays in cooler climates the Quonset type (Fig. 4) construction is more popular than 'A' shaped roof structure. Such houses are inexpensive, usually consisting of a framework of pipes, with or without any post or pole inside.

The climate control mechanisms of these structures are generally more specific and permanent, thus obviously expensive. The arrangements of air circulation, ventilation, cooling, heating, etc are more efficient than those of greenhouses meant for cultivation. Arrangement of light and shade, to modify the requirement of photosynthetic light and/or modify the day-length is frequently associated with such type of greenhouses. Pest and disease measuring and control mechanism are closely monitored in these greenhouses.

2.5.2. Greenhouse for cultivation of crop

Such greenhouses are relatively big. It may be a single greenhouse or a cluster of a few such single greenhouses. It may be of single-span or multi-span, low-cost or high-cost, covered with film or net. It is finalized depending upon the situation of climate, land, crop and availability of raw materials. However, in all the cases it must be related to crop production vis-a-vis the commercial viability.

As stated earlier, greenhouse meant for crop production is primarily to manipulate and manage the natural effect of light, temperature, humidity and precipitation in favour of desired growth of the crop. On the other hand, due to this, some microclimate related problems crop up under greenhouse condition. These problems are required to be taken care in such a way that they do not deter the growth of the crop. These crop-microclimate problems in greenhouse are related to crop-temperature, carbon-dioxide and inside-humidity.

For cultivation purposes, type of crop is the most important aspect for designing or selecting a greenhouse. In general flowers, vegetables and some fruits are cultivated under greenhouse. So the specifications of such greenhouses are determined on the basis of the specific type and variety of crop.

There are commercially acclaimed greenhouse-flower crops such as rose, gerbera, carnation, Anthurium, Orchids, Lilium, etc., which are grown as single-greenhouse-crop due to its perennial nature. Each of these flower crops requires completely different technology to grow commercially under greenhouse. Furthermore, depending upon the variety or the flower the necessary technology has to be adjusted.

These greenhouses shall have specific irrigation and fertigation arrangements. Pest- control measures like netting in the ventilators and openings to avoid attack of pests and vectors (pests that spread disease), apart from other stationed (traps) and mobile (sprayers etc) devices and tools. However, these insect-proof nets reduce the efficiency of natural ventilation (see chapter 4.3.4). To compensate this, the area of ventilation shall be adjusted accordingly and/or fan-tube exhaust system may be introduced in the greenhouse.

The same principle is applicable for all commercial greenhouse vegetable crops. The common greenhouse-vegetables are Tomato, Cucumber, capsicum, Egg-plant, lettuce, and different leafy vegetables. Vegetables are grown throughout the year inside the greenhouses, either as a single crop or with a suitable crop-rotation. Generally the life span of different greenhouse-vegetables ranges from 45 days to 1 year. Thus such greenhouse structures are more flexible in respect of design, ventilation, temperature and radiation control mechanisms to accommodate different kind of crops in a single unit.

According to the season and requirement of crop, the type of covering material is determined. Plastic film, plastic shade-net or both in combination maybe used as covering material of such commercial greenhouse.

The galvanized steel or locally available wood/bamboo may be used as the basic structural material. Depending upon the type of covering, wind flow, selected crops and economic capacity of the promoter, the structural material and corresponding technology may be chosen.

However, the life of plastic covering material (3-5 yrs) is always less than the life of the structure, and this provides a good scope of thorough maintenance of the structure after every 3-5 years. The life of local made wood/bamboo structures is less in comparison and is about 3 to 4 years years, similar to the life of plastic covering materials. Actually lives of plastic cover of such poorly engineered structures are also less due to wear & tear.

Bamboo Greenhouse Technology: Proper selection of bamboo, adequate curing & treatment process, scientific design and erection technique, proper fixing of covering materials, and best possible foundation/grouting system of such structure can create a commercial bamboo-greenhouse which can compete with steel structure and last for more than 10 years with due and scheduled (light & heavy) maintenance. Such eco-friendly and energy efficient greenhouse can be used for commercial cultivation of flower, vegetables, and fruits which has been discussed earlier. (see Chapter 5 for detail of innovated Bamboo Greenhouse Technology)

3

Factors for Selection of Greenhouse Technology

3.1. Non-Climatic Factors

Apart from climate several other non-climatic factors has to be considered before selection of proper greenhouse technology, which are listed in the following text.

1. Location and selection of site: The ideal location for greenhouse cultivation is an area having moderate climate with high sunlight, low humidity and good natural drainage. It is better to avoid storm prone areas or it is a must to take necessary protection, like wind-break and shelter belt, in such areas. However, selection of site, where the greenhouse will be constructed, is the first consideration to select the proper greenhouse technology. The proper selection of site will also depend upon various non-technical factors, which are listed below.

- Socio-economic and agrarian situation of the locality.
- Administrative and legal position of the area.
- Access to transport, power, manpower, and water.
- Geographical status of the area.
- Possibilities of any extreme climatic aberrations like flood, drought, storm etc.

Big trees or hillock may be considered in a proper place that will act as windbreak for the greenhouse. However closely placed such tall barriers should not be allowed to avoid shadow and normal wind flow. High altitudes in warmer area provide better range of temperature for crop growth; only the fall of night temperature is to be managed carefully. The site should have sufficient room for further expansion of greenhouse and associated structures.

2. Topography: Actual place/site of the greenhouse and sometime the floor area depends on the topography. The slope of land and the width of terrace frequently determine the size or planning of the greenhouse. The direction or face of the slope shall be taken into consideration at the time of selection of the site. For example south facing slope or presence of a high land or hillock in the north or northwest side is always preferred. The topography and slope shall be utilized to facilitate the natural ventilation and drainage of the greenhouse.

3. Vegetation: In sub-tropics, it is always wise to select a site with natural wind-break in the north and northwest in the form of moderately dense tall vegetation or tree line. However, such tree lines in close proximity shall be avoided in the south and southeast so that no shadow is cast from that side and air movement in and around the greenhouse is not restricted. In regions where snow is expected, trees should be 100ft away in order to keep drifts back from the greenhouse.

4. Accessibility: An easily accessible site shall be selected for greenhouse for easy incoming of agronomic and other inputs and smooth outgoing of produces which are generally highly degradable or perishable. This accessibility means (i) easy road transport directly to the market or to the port (air or ship) for exporting or long distance marketing, (ii) easy power source for continuous supply of electricity, (iii) availability of communication facilities like post, phone, fax, internet etc.

5. Source of Water: Nowadays the most precious and uncertain substance is water. Sufficient water for irrigation purpose throughout the year is not available in most part of the globe. Crop cultivation under greenhouse is dependent on 100% irrigation (except shade/net house). So availability of water in respect of quality, quantity and round the year supply has to be ascertained before finalization of the site. At the same time proper storage facility/structure for water collected from source or harvested from greenhouse has to be considered at the time of site selection.

Apart from scarcity of water, quality of irrigation water is an important factor for a greenhouse site. There are several cases in US where the greenhouses located in coastal and river bottom regions have been compelled to shift to new locations to obtain water of suitable quality. The cost of removing ions such as sodium and chloride can be prohibitive, but failure to do so results in plant injury.

6. Commercial Viability: Market of the product is the basic parameter to make a greenhouse project commercially viable. Depending upon the market price of the crop or plant grown under greenhouse (both present and future of the local and outside/export market), the magnitude of investment and selection of technology of a greenhouse shall be determined. Actual sale realization of the produce against the fixed cost associated with a greenhouse will finally decide the commercial viability of that project, keeping in mind the requirement of different recurring costs.

The 100% high-end/export market oriented large projects having the capacity to withstand financial loss during the initial years, can go for big investment and most sophisticated technology. These projects may finally prove to be commercially viable after a gestation period.

The other commercial projects for selected produce generally target both the local and distant market. These greenhouses shall compromise with cost in respect of construction, maintenance and technology to start with as a commercially viable project. There is ample scope to expand such projects in a bigger way, in due course, with minimal changeover costs. This will be the best strategy for making a commercially viable greenhouse technology for cultivation of suitable crops in a new area.

Commonly, in greenhouse cultivation, people do not want to compromise with quality of the produce and yield of the crop but, frequently it is dependent on climate and market. This is actually the reason of failure of many greenhouse projects, particularly in new areas. To make such type of projects commercially viable some compromise with both quality and quantity of crop as well as greenhouse technology is needed to be adopted.

However, it is a big question, how one should reduce the cost of a greenhouse and to what extent the technology can be compromised, while achieving more or less the same target. It is basically the flexibility of the greenhouse technology that have the capacity to guide or accommodate the changes one desires to make in a greenhouse project. The reduction of cost may be done either by reducing the cost of super structure in accordance to the proper calculation of load or by using locally available construction materials like wood, bamboo, plastic pipes etc. Another aspect to reduce cost is to adopt easy, efficient, but effective climate control mechanism utilizing natural environment. If one judicially and technically employ all these options of cost reduction, with proper combination, then he can make a more commercially viable greenhouse project, particularly in a new area.

So to adopt this flexibility, one shall have or acquire a very clear idea about the basics of greenhouse technology.

7. Availability of labour and other inputs: Apart from the export/high-end market oriented big projects, all other projects should identify the present and future labour demand, mainly skilled, and its supply. Next the supply of suitable construction materials that is cheap and easy to avail. In maintenance of greenhouse the cost and availability of labour and initial cost of greenhouse are directly proportionate. High labour cost increases the cost of greenhouse. To reduce the maintenance costs one may increase the durability of structural material (as in case of high tech/cost greenhouse) or vice versa (as in case of low tech/cost greenhouse). Most of the developing and under developed countries are situated in tropical regions where the flexibility of greenhouse technology is needed more due to absence of extreme climates and abundance of cheap labour.

3.2. Climatic Factors

Climate is a combination of above ground environmental factors e.g. temperature, moisture, sunlight and air which characterizes a region. Some regions, on an average, have very low level of moisture and are said to be arid or dry, while some other area have high moisture status in air, due to higher rainfall and profuse surface water bodies, are called humid region. Some areas exhibit both these characters in two different seasons. Similarly depending upon the status of average temperature, cool and hot regions are identified separately.

The season-wise average day-night temperature of the locality and difference between these two should be considered carefully at the time of selection of a greenhouse.

The immediate environment around the crop is called crop-microclimate. In addition to ambient weather, the evaporation and transpiration around the crop plays a great role in crop-microclimate that has a close relation with crop growth.

The facilities provided to the crop through the procedures of growing should be best arranged to optimize the response of crops against some fundamental factors like radiation, temperature, gases, humidity, precipitation and wind velocity. A well-planned greenhouse will provide such facilities in proper form to the crops cultivated inside that greenhouse. Among the above climatic factors precipitation is less important in cultivation under greenhouse. The plants are protected fully from rainfall under poly-greenhouse and the plants receive water only through suitable irrigation system. Thus the other climatic factors are described below in detail that will help to plan a proper greenhouse technology suitable to a specific area.

3.2.1. Solar radiation

At the edge of the earth's atmosphere, about 10 km above the earth's surface, the average solar radiation is 1367 W/m^2. Thus the total heat received by earth fron sun is 1.75 X 10^{14} KW. While passing through atmosphere, intensity of solar radiation is reduced by reflection, absorption, and scattering, so only a part of it reaches the earth's surface.

Total Radiation: The effect of solar radiation that reaches the earth's surface changes with latitude, altitude, and duration of the daytime and also with clouds/seasons. The average solar radiation of very low mean values is received by the earth's surface at the poles up to 20° latitude due to angular sunrays. In the equatorial zone, where the solar radiation falls straight, the mean annual solar radiation in between latitudes 25^0 south to 30^0 north is highest and remains almost constant (Fig. 5).

This high amount of solar radiation in equatorial area (sub-tropic and tropic region) depicts the high temperature prevailing there. This high temperature also affects the rainfall and storm pattern and subsequent high humidity.

For production of crop or growth of plants, it is important to know the mean daily sum of solar radiation energy for each month. This will provide the necessary information of difference of the mean daily sum of solar radiation between any two months. This information will help to plan proper greenhouse technology that can manipulate this solar radiation in favour of crop grown under greenhouse by controlling light intensity (irradiance), light duration, and light quality.

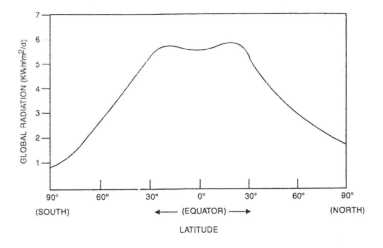

Fig. 5: Variation of annual means of global radiation depending on the latitude

Difference of radiation between summer & winter: On the other hand, with the increase of latitude, in both the hemisphere, the difference of mean daily solar radiation between summer and winter increases gradually (Fig. 6). In equatorial region this difference of radiation is less and sometime almost same.

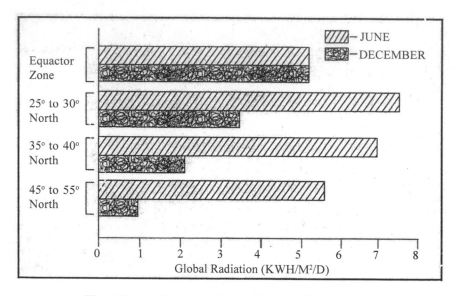

Fig. 6. Mean daily sum of global radiation in different latitude

The day length is also related with the mean daily sum of solar radiation received by the earth surface; more the length of the day more is the total radiation received. However, the production of crop or growth of plant decreases almost linearly with decreased radiation. Photosynthesis, the production mechanism of plant, will reduce with reducing solar radiation values. Growth of plant comes to a halt at a specific point called 'comparison point', i.e. 10 W/m^2 light power for 10 hour photoperiod (100 Wh/m^2/day).

Low numbers of daily light hours, low angle of sunrays, several cloudy and overcast days in growing season of the crop are the reasons for poor production capacity of that area. In case of greenhouse, low transmissivity of covering material due to quality of the film, dirt or otherwise are the reasons for further reduction of solar radiation received by a greenhouse crop. For a particular area these factors shall be considered to conceive the design of a greenhouse and its cropping system.

The solar radiation at noon is always high in comparison to the other parts of the day. However, in some places the radiation received by a plant in summer, throughout the day, can be too high despite of shorter day length. A minimum daily radiation of 500-1000 Wh/m^2/day is the limit for sufficient growth of a plant. In tropical region varieties suitable to shorter day length must be chosen for proper use of the high mean daily radiation.

BOX - 4

Irradiance: It is the relative amount of light as measured by radiant energy per unit area. Irradiance, intensity and photon flux all measures the amount of light differently.

The process of photosynthesis relies on the number of photons intercepted by the leaf area exposed to sun and not by radiant energy or intensity of sunshine received by the plant as a whole. The number of photons i.e. photosynthetic photon flux (PPF) should be used to relate plant growth, which is measured by quantum sensor and expressed as micromoles of photon per unit area per time ($\mu mol/m^2/S$). This can also be measured as watts per square meter (W/m^2) with pyranometric (radiometric) sensor. Some people still measure light intensity with photometric sensor, which determines foot-candles or lux (1 foot candle = 10.8 lux). A photometric sensor is relatively insensitive to wavelengths that are important for plant growth. So quantum or radiometric sensors should be used for greenhouse.

Photoperiodism: In general the higher or vascular plants are classified as long day, short-day and day-neutral based on the effect of photoperiod in initiation of reproductive growth. Long- day plants will flower when the critical photoperiod of light equals or exceeds the normal day length and short-day plants (chrysanthemums) flower when critical photoperiod does not exceed the same. Gamer & Allard, who discovered the photoperiodism of plant, demonstrated that the dark period is most crucial in initiation of reproductive growth.

Photoperiod is normally extended in greenhouse under short-day condition by lighting with incandescent lights or high intensity discharge light. On the other hand the photoperiod can be shortened under long day condition by covering the crop with black cloth or plastic in the greenhouse.

Light quality: Light quality is judged by colour, which corresponds to a specific range of wavelengths. However wavelength is the main criteria of the quality of light for plants. Plants i.e. crops responses differently in different coloured light (Fig. 7). For example far-red light can promote bulb formation in long-day plants, such as onion *(Allium cepa)*. Blue light enhances *in vitro* bud regeneration of tomato (*Solanum* spp.). Ultraviolet (UV) light has short wavelength, below 400 nm, which is not visible to human eye and in larger quantity is harmful to plants. Visible or white light ranges between 400 to 700 nm (Fig. 8). The impaired energy occurs at longer wavelengths and is not involved in plant growth. Light quality can be commercially manipulated through greenhouse spectral filters, greenhouse coverings, and varying supplementary light sources (detail in chapter-6.1.3.). This manipulation plays an important role in shoot development.

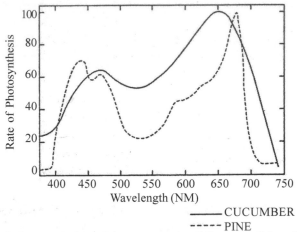

Fig. 7. Rates of photosynthetic activity occur under different qualities of light between the ultraviolet wavelength of 350nm and the farred wavelength of 750 nm. *(From The Electric Council, Growelectric Handbook No. 1. London, England)*

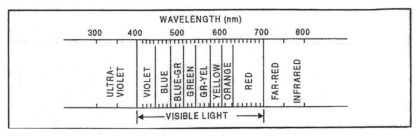

Fig. 8. Types of radiant energy have wavelengths of 300 to 800 nm. Visible light occurs in between the range of 400 to 700 nm.

BOX - 5

Photosynthetic Active Radiation (PAR): PAR designates the spectral range (wave band) of solar radiation from 400 to 700 nanometers that photosynthetic organisms are able to use in the process of photosynthesis. However, photosynthesis is a quantum process and the chemical reactions of photosynthesis are more dependent on the number of photons than the energy contained in the photons. Thus, the **PAR** is expressed as the amount of energy by which the instantaneous light incidents upon a (plant) surface, and is measured by micromole per square meter per second **(μmol/m^2/s)**. Here 1 mol is $6{,}02 \times 10^{23}$ photons and 1 micromol (μmol) is 1-millionth of a mol. On the other hand, daily light integral (DLI) is the accumulation of the entire PAR received during a day. The unit for cumulative light or daily light integral (DLI) is – moles per square meter per day (or mol/m^2/day).

Since plant photosynthesis is the main use of supplemental light in the green-house, light sources are often rated on the amount of Photosynthetic Active Radiation (PAR) delivered to the plant surface.

BOX - 6

Supplementary photosynthetic lighting in greenhouse: In greenhouse low rate of photosynthesis in cloudy short-day-length period may be over-come by supplement photosynthetic lighting. High-pressure sodium (HPS) vapour lamps emit more photo-synthetically active radiation (PAR), in com-parison to other type of lamps, for each input watt of electricity. This also provides heat and less shading to the crop. The installation should provide a minimum of 65 μmoI/m^2/S or 13 W/m^2 PAR at the plant level with a 16 hour photoperiod (see chapter 6.1.3). Different aspects of solar radiation in sunny day and cloudy day are given by (Thimijan R. W. and R. D. Heirs 1983).

Table 3: A comparison among energy, radiation and illumination of sunlight occurred in sunny day, cloudy day and supplement light.

Sun Light Condition	Energy (PPF) [μmol/m^2/s]	Radiation (irradiance) [W/m^2]	Illumination (light intensity) [lux]
Full sunlight	2000	450	108000
Heavy cloud	'60	15	3200
Supplement light at 2m height (metal halid, 400w)	19	4	1330*

* Different artificial light with capacity to supplement illuminations up to 2000 to 10000 lux are commercially used.

3.2.2. Temperature

For greenhouse technology the temperature, one of the climatic factors, should be considered in two different ways. First, the outside/ambient temperature on which the structural design of a greenhouse depends. Second is the inside/greenhouse temperature, which is modified in favour of the crop, grown under the greenhouse.

3.2.2.1. Outside or Natural Temperature

Season that is governed by latitude, altitude, distance from sea, wind condition and finally the intensity of solar radiation control the course of temperature prevailing in a specific location. Therefore it is difficult to make any generalization about temperature status of a vast region. Even two places separated by a hillock may exhibit different temperature pattern, though in a minor scale.

Effect of Latitude: As the effect of solar irradiation and season on earth changes with changes of latitude, the phenomenon of temperature along the latitude from equator towards pole of both the hemisphere is changed following a general pattern e.g. the average temperature reduces with increase of latitude. This pattern is closely related with the status of solar radiation along the latitude (Figure - 4). However, the daily temperature increases initially from equator to approximately 25^0 towards pole in both hemisphere and then reduces towards pole.

The mean of the maximum temperatures in the warmest and coldest months of the year at Equator ($0°$), $27°$ latitude and $50°$ latitude are 32.5, 37 and 23°C, ($T_{max}W$) and are 31, 21 and 1° C ($T_{max}C$) respectively. The result will be similar in both the hemispheres if the local factors like altitude, distance from sea etc. remains the same. On the other hand, the mean minimum temperatures in warmest and coldest months of the year at Equator, $27°$ and $50°$ latitude are 23.6, 26.5 and 12.4°C, in warmest month ($T_{min}W$), and are 22, 10.8 and (-) 5°C in coldest month (T_{min} C) respectively.

Table 4: Status of temperature along the latitude of the earth.

Types of temperature	Temperature in °C		
	0° latitude	27° latitude	50° latitude
1. $T_{max}W$	32.5	37	23
2. T_{max} C	31	21	1
3. Extreme max. Temperature	35.5	50.0	37.5
4. $T_{min}W$	23.6	26.5	12.4
5. $T_{min}C$	22	10.8	-5
6. Extreme min temp.	19	(-) 4	(-) 46

The above table clearly indicates that, in general, the equatorial area almost has no requirement of temperature controlling mechanism for cultivation. Only a protective roof is sufficient enough to control the rainwater and very high intensity of solar radiation. So shelter type greenhouse structure is good for cultivation if local climate is favourable for crop growth. The higher latitude areas have the problem of low temperature at night particularly in coldest months and increasing the inside temperature of the greenhouse to a certain extent is required.

The greenhouses in the climatic zones around 20° to 30° north and south latitudes may require both the cooling and heating mechanism to cultivate crops throughout the year to avoid problems at extreme temperatures and danger of frost. Of these two mechanisms cooling is more important to reduce the temperature sufficiently in hottest months. The 'Hot-house' mechanism, utilizing solely the concept of 'greenhouse effect', to increase the night temperature in cooler months may be sufficient enough for crop production. If necessary, separate heating facility may be arranged for such cool nights.

Effect of Day and Night Temperature: The difference between day and night temperatures [DIF — difference between day (DT) and night (NT) temperature] is a major environmental factor affecting crop growth, but the mechanisms are not fully understood.

Study by S M P Carvalho & others shows that the response of chrysanthemum's final internodes-length to temperature is strongly related to DIF. However, it also shows that this response is simply the result of independent and opposite effects of day and night temperatures. Finally, It is concluded that although the DIF concept does not have a biological meaning, it can be a good predictor of final internodes-length of chrysanthemum within a temperature range of 16–28 °C.

In case of growth of root of tomato, positive DIFs have higher positive influence than negative DIFs, and 6°C DIF is best for greenhouse tomato growth (Zaiqiang Yang *et al.*)

It appears that the difference of day and night temperature affects different biological activity of the crop, particularly flower initiation, fruit setting and vegetative & root growth. Thus, night temperature has an important role on growth of different crops, which can be manipulated under greenhouse condition.

Apart from all these climatic data, the local variation in micro-climate has to be examined properly to establish a greenhouse.

Effect of Altitude, Distance from sea, etc: The temperature can change with altitude, distance from sea, wind direction, and cloud condition. These geo-climatic conditions influence the temperature in the following way:

- Temperature drops with increase in altitude above sea level. On an average for every 1000ft increase of altitude the average temperature is reduced by 1°C. However, with the higher altitude the night temperature reduces more than that of day temperature (Fig. 9).

- The amplitude of temperature significantly affected by the increasing distance from the seas. The average temperature always increases nearer to sea between the two places with same latitude (Fig. 9).

- In winter the advancing cold air towards equator from poles decreases the temperature further.

- The cloudy overcast sky reduces warming up in day time and also reduces cooling in night, thus the range of temperature (Maximum & Minimum or average day & night temperature) remains more or less even.

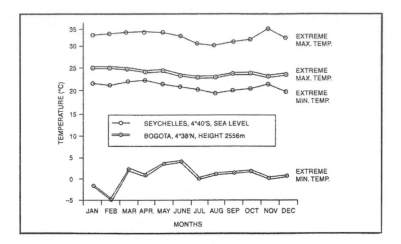

Fig. 9. Variations of extreme maximum and minimum monthly temperatures having same amount of radiation and different altitude as well as distance from sea.

The figure-8 is a typical example of the effect of both altitude and distance from sea on temperature. The graph exhibits the extreme maximum and minimum temperatures attained in different months of a year of Bogota, Colombia situated at 4°38' north latitude, 2556 m altitude and about 350 km away from sea and of Seychelles, an island of Indian Ocean nearest to Africa, situated at 4⁰40' south latitude, almost at sea level and surrounded by sea.

The interesting information available from the figure is that despite having the same latitude, the 'extreme maximum temperature' in a year varies between 30 to 35°C and 23 to 25°C at Seychelles and Bogota respectively. The 'extreme minimum temperature' in the same period between these two places differs more widely and the figures are 20 to 22.5°C and -5 to 9°C respectively. This is primarily due to the combine effect of high altitude and distance from sea.

So, while having the same solar radiation the cultivation of vegetable crop throughout the year is very much possible in simple greenhouse at Seychelles, in Bogota better quality greenhouse having the capacity to increase night temperature is essential for cultivation of vegetable throughout the year.

Importance of extreme minimum and maximum temperature over the mean maximum and minimum temperature: -

Even month-wise mean minimum and maximum temperatures for a few years (may be 10 years) does not provide the specific information for selection of specific greenhouse technology. This will provide general information to conceive a basic idea about the type of greenhouse technology to be adopted. But to specify the technology, extreme monthly minimum and maximum temperature figures, for round the year, for last few years (may be 10 years) is essential.

The fluctuation of outside temperature is important to plan the climatic control mechanism or system of a greenhouse.

3.2.2.2. Inside Greenhouse Temperature

This discussion is necessary and important in respect of crops grown under any greenhouse. However, generally the daytime inside temperature of a fully closed greenhouse is always more (minimum 20%) than that of outside. Now the openings provided in the greenhouse for ventilation can reduce such hike of temperature to an extent, depending upon the ventilation efficiency of the greenhouse. Proper natural ventilation system can equalize the temperature of outside and inside, particularly in smaller greenhouses, if the design of the greenhouse is done properly. Some artificial forced ventilation (using fans) can reduce the inside temperature further in daytime. However, with more opening areas or windows in greenhouse walls, it will be very difficult to use exhaust system. For that situation an innovative exhaust system (fan-tube-exhaust) is required to install.

When cooling is required: For further reduction of inside temperature a proper cooling mechanism is required. In general, in the area of very high solar radiation, the temporary system of shading will help to reduce inside temperature to the extent of 4-5° C. In dry and hot areas, evaporative cooling mechanism can be used for cooling the greenhouse to a great extent (10 to 15° C) in daytime (detail in chapter 6.2.1.).

When heating is required: As night approaches, the outside temperature gradually drops with withdrawal of solar radiation. If the mean night temperature goes below the level required for optimum biological activity (12°C), the night temperature inside the greenhouse can be increased by closing the ventilators of greenhouses at afternoon. This closed condition of greenhouse can retain the day time high temperature for the night. If the mean daily outside temperature goes below 6°C, for more than two consecutive days and such event happens several times in each year, then separate heating arrangement is required to increase the inside temperature at night.

Thus one of the main purposes of greenhouse is, to provide, as best as possible, an average daily temperature, with minimum fluctuation, suitable to the crop grown under it.

3.2.3. Precipitation

It is necessary to have the information about the characteristics of rainfall, snowfall etc. to determine the structural design of a greenhouse in that region. These characteristics are quantity, seasonality, intensity and frequency of average single rain.

Though rain-water is a vital element for growth of crop in open-field, in poly-greenhouse rain is insignificant as the crop grows under cover is irrigated exclusively. However variability of different characters of rain from year to year is important for crop under different type of net-house.

However climatic data about occurrence and intensity (kg/m^2) of Hail or hailstorm is a very vital aspect of greenhouse designing. Depending upon that information the strength of cladding material is to be selected. The places with frequent occurrence of high intensity hail-storm (more than $25kg/m^2$), one may use watertight, UV-resistant, and anti drop, woven PE-sheet instead of normal PE-film.

BOX- 7

Atmospheric Precipitation

When relatively hot air close to earth surface ascends it expands due to low pressure of the upper region and gets cooled down. As a result, the water vapour present in it condenses, which leads to the formation of water drop-lets or small ice crystals, thus making cloud or mist. When water droplets coalesce in cloud and grow bigger or heavy enough, they fall to earth as precipitation.

The common forms of precipitation are rain, snow, drizzle and hail. In case of **rain** the size of water drops/droplets is 0.5 mm to 6.0 mm in diameter. The larger than 6mm tend to break up into smaller drops during their fall. Drizzle is a form of rain when numerous water droplets of size less than 0.5 mm with an intensity less than 1 mm/hour falls on ground surface. Snow consists of ice crystals that usually combine to form flakes. Fresh snowfall has an initial density varying from 0.06 to 0.15 g/cm^3, with an average value of 0.1g/cm^3 (i.e. 10 mm snow = 1 mm rain). In case of Hail or hailstorm the irregular pellets or lumps of frozen rain, of size greater than 8mm, occurs as precipitation. This results when upward air currents are very strong, blow-ing the raindrops up to freezing heights.

Depending upon the intensity the rainfall is classified as light, moderate and **heavy**. Rainfall intensity trace to 2.5 mm/hr, 2.5 to 7.5 mm/hr and more than 7.5 mm/hr are generally termed as light, moderate and heavy respec-tively. In the tropics the layer of cloud is up to 10 km thick.

The rain with intensity of 30 to 60 mm / hr or higher, is called very heavy rain or **shower** providing a lot of water in a short period.

Excess water in root zone, and high intensity rainfall at seedling/young stage damage the crop in open field condition to a great extent. This situation in rainy season restricts the growth of many crops almost for the entire season. In many areas the period stretches from 3 to 5 months. Due to soggy saturated condition of root-zone soil the crop growth is hampered and the optimal yield cannot be achieved. This situation commonly prevails in low and fertile land situation, which are otherwise denoted as best agriculture land. To avoid this problem properly designed greenhouse is an essential tool for such areas. In such greenhouse design of gutter is an important aspect.

Drainage: It is an important aspect that is related to precipitation, particularly rainfall. In such regions and also in other regions one of the main jobs of the greenhouse is to protect the root zone of the crop from excessive water. While planning the drainage system around the greenhouse is important, draining out the precipitation water away from its vicinity through a properly designed drainage system shall be considered carefully. Depending on the duration and characteristics of rainfall the drainage system should be designed. Besides, rain-water should not enter the greenhouse through seepage in any circumstance.

In cool climate, where snowfall is common in winter, the greenhouse and its drainage system should be designed in a way that the snow easily glides away from the roof and the subsequent melted water will be drained out from the vicinity of the greenhouse.

3.2.4. Humidity

The importance of air humidity for designing greenhouse for crop production is frequently underestimated. Both the humidity of air outside and inside of a greenhouse shall be meticulously evaluated and considered while planning a greenhouse. Furthermore, both absolute humidity and relative humidity has to be properly measured and shall be considered with due importance. Plant growth and production will slow down or stop when the humidity in air is lower than 30% or higher than 90%.

The following two reasons, in general, may be decisive for a deviation of relative humidity in the greenhouse:

1. The varying exchange condition of air volume in respect of water vapour at an equal rate of water vapour supply by way of evapotranspiration.
2. The varying inside temperature condition of the greenhouse influences the relative humidity considerably. The high inside temperature reduces the relative humidity of inside air of the greenhouse, but due to high transpiration the absolute humidity increases accordingly.

3.2.4.1. What is humidity?

Humidity is the water vapour present in the air. Status of humidity is generally considered and measured in different ways in relation to temperature viz. relative humidity (RH), absolute humidity, vapour pressure deficit (VPD) and dew point.

Relative humidity: It is the water vapour present in percentage of maximum water vapour that can be retained by air at a given temperature. It is expressed as the ratio of water vapour actually present in the air and maximum water vapour that can be retained by that air at a given temperature. For example per cubic meter of air at 20°C can hold 17.3g water vapour in maximum (100%). So if it holds 13g/m³, then the relative humidity is 75%. The relative humidity falls with increase in temperature and rises with decrease in temperature. Now, the most problematic situation is higher relative humidity at higher temperature.

Absolute Humidity: It is the actual weight of water vapour per cubic meter of air and expressed as gram/m³. Conversion of absolute humidity to relative humidity can be done only when the temperature is known. For example 15g/m³ at 17.5°C equals to 100% relative humidity while with same absolute humidity at 25°C the relative humidity equals to 65% and at 30°C it equals to 50%. Thus, under greenhouse, when absolute humidity is very high at higher temperature, it is really difficult to control.

Vapour Pressure Deficit: VPD indicates 'drying effect' of the air. It is the relation between vapour pressure of a particular gas, here the water vapour, and the normal air pressure (pressure of all gases in the air together) that is expressed as millibar or kilo-pascal (KPa). It normally ranges from 10 to 50 millibar or from 1 to 4 KPa.

Dew-point: It is a point of temperature and absolute humidity below which the excess water vapour in air condenses. For example air with 10g/m³ and 20 g/m³ has dew point of 11°C and 22.5°C respectively. That means, under greenhouse condition, when dew-point arrived, the relative humidity will reach 100%, and any further drop of temperature reduces the maximum holding capacity of water vapour of that air. The excess water vapour condenses and precipitates as dew on the crop, which can invite or spread the fungal infection.

Effect of Relative Humidity: We understand that RH 95% at 20°C means 16.5 gm/m³ absolute humidity (Table 5). In theory RH below 100% means that there is no condensation, but the temperature throughout the greenhouse may not be equal and in few corner spots the RH may be 100%. As a result condensation takes place on the leaves. Thus the spores of *Botrytise* or other fungal disease starts growing. So any greenhouse owner should know the RH reading and its variation inside, and monitor it regularly.

Relative humidity affects leaf area development and stomatal conductance thereby interfering with the photosynthesis and dry matter production (Jolliet, 1994).

High levels of relative humidity can lead to yield loss especially for tomato crop (Bakker, 1990).

Effect of Vapour Pressure Deficit (VPD): VPD is the difference between the actual and the maximum vapour pressure of the given air at any particular temperature. T

At every such situation there is minimum and maximum pressure of water vapour, which ranges from 1 Kpa (very wet) to 3 Kpa (very dry) and for each temperature there is a specific vapour pressure beyond which it condenses.

Table 5: Comparison of relative humidity, absolute humidity and VPD.

Temp.	10°C		15°C		20°C		25°C		30°C	
RH	AH	VPD	AH	VPD	AH	VPD	AH	VPD	AH	VPD
(%)	(g/m³)	(Kpa)	(g/m³)	(Kpa)	(g/m³)	(Kpa)	(g/m³)	(Kpa)	(g/m³)i.V>.	(Kpa)
100	9.42	0	12.86	0	17.30	0	23.09	0	30.43	0
95	8.94	0.06	12.21	0.09	16.47	0.12	21.94	0.16	28.91	0.21
90	8.41	0.12	11.57	0.17	15.60	0.23	20.79	0.32	27.39	0.42
85	8.00	0.18	10.93	0.26	14.73	0.35	19.63	0.48	25.87	0.64
80	7.53	0.25	10.28	0.34	13.87	0.47	18.84	0.63	24.34	0.85
75	7.06	0.31	9.64	0.43	13.00	0.59	17.32	0.79	22.82	1.06
70	6.59	0.37	9.00	0.51	12.13	0.70	16.17	0.95	21.30	1.27
60	5.65	0.49	7.71	0.68	10.40	0.94	13.86	1.27	18.26	1.70
50	4.21	0.61	6.43	0.85	8.67	1.17	11.55	1.59	15.22	2.12
40	3.77	0.74	5.14	1.02	6.93	1.41	9.24	1.90	12.17	2.55
30	2.82	0.86	3.86	1.20	5.20	1.64	6.93	2.22	9.13	2.97

3.2.4.2. Measurement of humidity

Humidity can be measured in four ways viz. (1) Relative Humidity (in %), (2) Absolute humidity (in g/m³), (3) moisture VPD (in Pa or KPa), and (4) Dew point (in °C). RH and VPD are the two major aspects chiefly considered to control humidity in sophisticated greenhouses.

RH (sometime dew point) is used to monitor infestation of disease and its control while VPD is the best indicator for transpiration. However, from any form of humidity measurement, one can compute the other forms of data to measure humidity.

At present three type of sensors with different measuring principles are commonly available and used in greenhouse: (1) Dry and Wet bulb instrument,

(2) Electronic humidity sensor or capacitive sensor and (3) hair-hygrometer (detail in chapter-3.3).

3.2.4.3. Air humidity outside greenhouse

The humidity in a greenhouse with proper ventilation system depends on the ambient/outside air humidity. Though, the crop density of greenhouse increases the humidity inside than that of outside, the higher temperature inside reduces the figure of RH. High relative humidity of ambient air of both cold and warm season affects the VPD.

In cold season when there is fog the RH is very high but the absolute humidity is very low because the cool air can hold only a small amount of water vapour. In warm season the outside RH is normally less in comparison. However, it can be very high due to absorption of water vapour coming from sea or rapid evaporation of rainwater/ water body and high rate of transpiration by plants.

In cold temperature rain does not necessarily cause high air humidity; in fact very cold rain droplets can reduce the absolute humidity due to condensation of air water vapour. In warmer temperature rain always increases the humidity of ambient air.

3.2.4.4. Air humidity inside greenhouse

The ambient absolute humidity is the starting point of the absolute humidity inside the greenhouse. Transpiring plants increase the absolute humidity of greenhouse air by adding water vapour, so it is generally higher inside than the outside air. However, the relative humidity can be higher or lower inside than the outside air, and this happens due to the temperature differences.

Movement of air inside the greenhouse is an important factor to control the humidity. Maintaining a gentle air-movement at the rate of $^1/_2$ to 1 meter/second around the plant is ideal. While there is some variation in optimum relative humidity levels in different crops, 75% to 85% is normally ideal. RH levels in excess of 95%, for example, can lead to reduction of transpiration pull of the crop, and that affect the intake of nutrients from soil leaving the plant vulnerable to a range of physiological disorders, thus reducing the yield.

3.2.5. Carbon dioxide (CO_2) status

Carbon is an essential plant nutrient and is present in the plant in a higher quantity than any other nutrients like Nitrogen, Phosphorous etc. About 40 percent of the dry plant matter is composed of carbon. This carbon is obtained only from the air in the form of CO_2, which constitute only 0.03 percent of air.

During the daytime, when the stomata of leaves are open, the CO_2 gas diffuses through these microscopic openings. Once inside, it moves within the cells where in presence of sun or solar energy and with the help of water it produces carbohydrates (sugars), the basic compound for growth of plant. The above mentioned process is called photosynthesis.

Relation between CO_2 uptake and photosynthesis of a plant: There is a very close relationship between CO_2 uptake and rate of photosynthesis. Both are affected similarly against light intensity and air-flow. Commonly the rate and total amount of photosynthesis is measured as well as expressed as "mm^3 of CO_2 /cm^2/h" and "mg of CO_2/100 cm^2/h" respectively.

At normal CO_2 level, maximum photosynthesis of individual leaf of many plant species takes place at light intensities around 21530 lux.

In open field, as the CO_2 percentage is constant in atmosphere, no significant change is possible. The plants are used to optimize this situation.

But in greenhouse the CO_2 status is completely different and the possibility of reduction of CO_2 in air, particularly in daytime, is very common. Artificial increase of CO_2 inside the greenhouse, which is called CO_2 injection, can be done to increase the growth rate of the crop. (The matter is discussed in detail in chapter 6.4.)

3.2.6. Wind speed and storms

3.2.6.1. Wind/Air velocity inside greenhouse

Under greenhouse condition, inside flow of air monitors several important aspects of microclimate that have great influence on the growth regulating processes of crop / plant viz. photosynthesis and transpiration.

Due to photosynthesis and respiration the microclimate around the leaf is changing regularly. In daytime CO_2 status of this microclimate reduces significantly due to increasing photosynthesis rate, which subsequently leads to gradual reduction in photosynthesis. At the same time the Relative humidity around the leaf increases due to transpiration eventually reducing the rate of transpiration. This ultimately affect the root uptake adversely.

The increase of air circulation inside a greenhouse induces the air movement in the leaf-microclimate, hence reduce the humidity and increase the CO_2 status around stomata. As a result rate of photosynthesis and root uptake increases, which induce more growth of the crop.

Gastra (1959) explored the relationship between CO_2 uptake, the rate of photosynthesis, and wind velocity of many greenhouse vegetables. He observed that, there is a continuing increase in rate of CO_2 uptake by leaf as wind velocity increases (Table 5). Wind velocity influences the resistance to CO_2 diffusion from air to leaf.

Table 6: Influence of wind velocity on maximum rate of photosynthesis of leaves in ambient air

Wind velocity (cm/s)	10	16	42	100	1000
Photosynthesis rate (mm³ CO_2 /cm²/h)	79	88	101	109	118

Where the air temperature is less than the leaf temperature, the plant temperature comes closer to air temperature by way of conduction of more heat away from plant (Fig. 10). If the air temperature is higher than plant temperature, the heat will flow into opposite direction, tending to raise plant temperature.

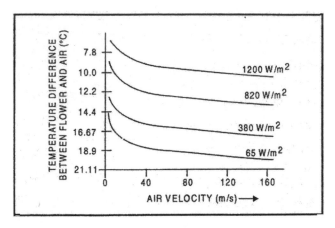

Fig. 10. Relation of temperature difference between plant (carnation flower) and air with Air Velocity in winter/cooler periods (Hanan, 1969). (When air temperature is higher than the plant, the heat will flow in opposite direction)

3.2.6.2. Wind/Air velocity outside greenhouse

High wind speed and storms can lead to the damage of a greenhouse and/or the inside crop. Thus, the greenhouses shall be built in such a way that they can resist the high-wind, particularly storms, otherwise the greenhouse may be damaged seriously or destroyed completely. If the high-wind penetrates the greenhouse through windows the crop, particularly the trellising crop, can be badly damaged.

Analyzing long-term (20 to 50 years) climatic data, the average wind speed can be estimated. Depending on that estimate the structural design should be made. It should be made in such a way that it can take care of the extreme adverse effect of wind/storm. In general 100 to 150 km/h wind speed is considered for greenhouse construction depending upon the area.

3.2.6.3. Storms

It is a fact that sometimes the gusty wind or storms are so strong that greenhouses cannot withstand the same. Such storms are generally developed in and around tropical and subtropical areas of the globe. The World Meteorological Organization (WMO) has adopted the following classification of tropical storms (Jackson 1977):

1. "Tropical depression": Low pressure, due to a hot & dry period in tropical and subtropical areas, enclosed within a few isobars, either lacking a marked air circulation or wind velocities; below 17 miles/second or 160 kmph is commonly termed as Tropical depression. Strong and very high wind-speed between 100 to 160 kmph for a reasonable period along with rain may cause havoc to crop and greenhouse.

2. "Tropical Storm": When a destroying wind travel across with a speed range of 17 to 32 miles/second (160 to 300 kmph) for a limited period we called it tropical storm.

Tropical storms usually develop within the zones lying north and south of the equator. The huge pressure gradient occurs due to temperature difference between sea and landmass of the equator. Due to evaporation from sea, the latent heat reduces the air temperature above it. On the other hand high intensity solar radiation increases the temperature of land mass and the air layer present in contact with it. This temperature difference in air generates the mass movement of air thus creating the storm or high velocity wind.

3. "Tropical Cyclones": Small diameter (some hundreds of kilometers), minimum surface pressure, very violent winds, torrential rain; usually containing a central "eye", with diameter some tens of kilometers, where there are light winds and lightly clouded skies.

The cyclones or hurricanes with wind velocities of more than 20 miles/second are predominant in areas within 26.5°C sea surface isotherm, which is placed around 20° of both side of equator. The Pacific Ocean and Indian Ocean areas have the highest frequency of tropical storm per year (Table 7).

The zones where hurricanes occur show heavy daily rainfall. In such regions, a high percentage of total rainfall occurs during hurricanes.

Table 7: Frequency of tropical storms (wind velocity more than 20m/second).

Location	Average no/year	% of worldwide total
1. Northeast Pacific	10	16
2. Northwest Pacific	22	36
3. Bay of Bengal	6	10
4. Arabian Sea	2	3
5. South Indian Ocean	6	10
6. Off northwest Australian coast	2	3
7. South Pacific	7	11
8. Northwest Atlantic	7	11

Jackson, 1997.

3.3. Instruments Required for Measuring Different Parameters of Greenhouse

Appropriate implementation and subsequent operation of greenhouse technology require study of different biotic and non-biotic data for taking any decision. For planning to implement this technology, analysis of long term information is essential. But implementation and subsequent operation need various non-biotic and biotic in-situ data through regular analysis, particularly for automation of climate control mechanisms. The important factor is that the instruments shall be small, portable and easy to operate. A brief note of these instruments is narrated below.

3.3.1 Measurement of non-biotic data

1. **Temperature:** (a) **Thermometer** made up of glass carrying mercury is highly accurate instrument to measure temperature (mostly air temperature) in the range of 0° to 110°C. This instrument works on the principal of expansion of liquid, usually mercury or alcohol coloured with die. (b) **Thermistor** is a thermally sensitive variable resistor made up of ceramic like semi-conducting material made from the oxides of copper, manganese, nickel, cobalt and lithium. It is a small device having wide operating range between -100° to 300°C and is capable to measure accurately up to 0.01°C. (c) **Thermocouple** is the most common electric method of temperature measurement that uses thermo-electric sensor. A thermoelectric electromotive force (e.m.f.) is generated in the device when the ends are maintained at different temperature, the magnitude of the e.f.m. being related to temperature difference. (d) **Digital electronic temperature indicator** is an electronic device that can directly display the temperature measured by thermocouple.

2. **Solar radiation**: (a) Sunshine recorder is a meteorological instrument that measure the duration, in hours of bright sunshine, during daytime. For operation of greenhouse, it is necessary to know the amount of solar radiation required. Thus, this instrument is secondary one for greenhouse cultivation. However, data of this instrument is useful for planning of a greenhouse. (b) Solarimeter or Pyranometer is an instrument to measure total hemispherical solar radiation by horizontally placing an object having sensitive surface exposed to sun. It can measure the diffuse solar radiation, (c) Actinometer or Pyrheliometer measures solar radiation (beam radiation) in normal incidence.

3. **Humidity**: (a) Dry and wet bulb thermometer is the common and simplest instrument that can measure humidity. It is simple and yet ingenious method to determine air humidity. It contains two thermometers placed side by side inside the greenhouse. One of the two measures the normal air temperature and the other, continuously wet, measures the temperature at wet condition. From the data, collected from these two thermometers, the RH may be determined manually. For automatic data recording these two thermometers may be placed inside a specially designed apparatus that generates the data continually and send to the computer that analyze and provide the humidity situation inside (Fig. 11). (b) Electronic (capacitive) sensor: Electronic sensor, mostly capacitive sensor produce a signal developed in the presence of water vapour in the air and transfers it to the computer or a digital system, which converts the signal to RH reading and display accordingly. This can become unreliable particularly in high humid conditions. After about 6 month it becomes inaccurate, and is required to be calibrated every month particularly in wet conditions. However, the recently developed digital instruments provide the perfect data for a longer period. (c) Hair Hygrometer: Human hair become longer as humidity increases and shorten when becomes dry. This property of hair has been used to form this instrument, which is quite cheap and is available in convenient pocket size shape. Such instruments, as digital hygrometers, are recommended for use in temperature range between 5° to 35°C and in the relative humidity range between 40 to 95 % of a greenhouse, (d) Humistor: It is the humidity resistor; consisting of two metal grids bounded to a sheet of plastic. As the relative humidity rises, the film becomes more conductive and the electric resistance of the grid is lowered. It can measure the large range of humidity of 5 to 99 %.

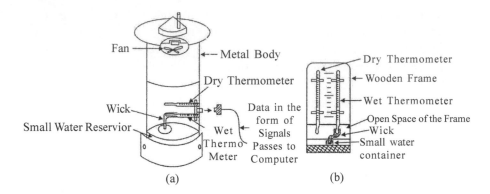

Fig. 11. Simple humidity measurement instrument (a) and Automatic humidity measurement instrument to feed computer (b) Using dry-buib and wet-bulb thermometers.

4. **Measurement of direction & speed of air flow:** Wind vane is a freely rotating arrow like structure installed 10 feet (3.05m) above the ground surface. The head of the arrow denotes the direction of wind and the geographical directions are prefixed with the wind vane. There should be no obstruction, and if any, that shall be away from the wind vane at a distance of at least 10 times of its height.

   Wind speed is measured normally by 'cup counter anemometer'. It is also a free rotating device consisting of three or four hemispherical cups. These cups are in motion when pressure difference develops between the two faces of the cups due to slightest flow of wind. The number of rotations made by the cups in a specific time can be counted by a counter and is directly calibrated into kilometers. The installation of the anemometer is similar to that of wind vane.

5. **pH:** For greenhouse technology measurement of pH is essential to know the acidity or alkalinity of soil and irrigation water. From crude measurement through litmus paper to most sophisticated pH-meters are used for greenhouse cultivation. The electrochemical pH-meter consists of electrodes, which actually measures the voltage to determine the pH of the tested sample solution. Null type and direct reading types are the two forms of such pH-meter. The null reading can be obtained quite readily using potentiometer circuit. To obtain accurate pH reading, the pH measuring instruments has to be calibrated.

6. **Soil Moisture:** The soil moisture content is commonly expressed as the *weight percentage* i.e. grams of water associated with 100 grams of dry soil. Moisture content is also sometimes expressed in *volume percentage*. It is the volume of soil water as a percentage of the volume of the soil sample. **Methods:** In the laboratory soil moisture content in samples can

be determined by methods like (a) *gravimetric method*, (b) *rapid moisture meter using calcium carbide*, (c) *soil moisture gauge*. In field condition or in-situ moisture content can be estimated by using (a) *resistance blocks*, (b) *Tensiometers* and (c) *neutron moisture meter*, (d) *Indicator plant* etc.

The resistance blocks i.e. the porous blocks made up of gypsum or nylon or fibreglass measures the electrical resistance of these blocks in different moisture regime. The resistance in the blocks is determined by the moisture content and in turn by the tension or suction of water in the nearby soil. The relationship between the resistance reading and the soil moisture percentage can be determined by calibration. This method gives reasonably accurate moisture readings over the range of 1 to 15 atmospheres suction. Tensiometers measure the tension with which water is held in soils. Here the water in the tensiometer equilibrates through a porous cup with adjacent soil water and that the suction in the soil is same as the suction in the potentiometer. Their range of usefulness is between 0 to 0.8 bar suction and is used for determining the need for irrigation. Through irrigation the moisture status of soil is kept near the field capacity. It can measure soil moisture of different depths by placing porous cup fitted at the end of the tensiometer at desire depths.

3.3.2 Measurement of biotic parameters

1. Photosynthesis through CO_2 uptake by plant: This system extends the boundaries of gas exchange and is carried out by providing a portable instrument. With typical CO_2 depletions of only 2 to 100 ppm (depending upon photosynthesis rate), the measurement is made very near ambient condition. High accuracy of the CO_2 analysis means that measurement of very small photosynthesis rates is possible. These instruments consist of a CO_2 analyzer, a system console and a sensor, housing with interchangeable leaf chambers. The improved type can calibrated for measurement of a range from 0 to 1500 ppm.

Some field ready, portable photosynthesis analyzer is available in the market to Measure photosynthesis, transpiration, stomatal conductance, and internal CO_2 concentration. This gas exchange measurement system is easy to operate in the lab or in the field/greenhouse, in any conditions. It comes ready to take ambient measurements of gas exchange right out of the case. Optional environmental modules allow to control CO_2, H_2O, temperature, and light, as well as measure chlorophyll fluorescence and photosynthesis rates simultaneously.

2. Leaf Area Index (LAI): It is a dimensionless quantity that characterizes plant canopies. It is defined as the one-sided green leaf area per unit ground surface area (LAI = **leaf area** / ground **area**, m^2 / m^2) in broadleaf canopies.

LAI is used to predict photosynthetic primary production as a reference tool for crop growth. This helps to determine leaf temperature turbulent transport, productivity, evapo-transpiration, interception of precipitation and solar radiation. An inverse exponential relation between LAI and light interception, which is linearly proportional to the primary production rate, has been established.

$$P = P_{max} (1 - e^{-c.LAI})$$

Where P_{max} designates the maximum primary production and c designates a crop-specific growth coefficient. This inverse exponential function is called the primary production function.

Direct measurement of leaf area on graph paper and subsequent relation of it with canopy structure is a tedious and intensive job, even for a small canopy.

Instruments like portable laser leaf area meter are used to track the growth and health of plants by measuring Leaf Area and LAI rapidly and non-destructively in any environment. Sometime the high-resolution laser scanner, data logger, and display are all enclosed in a durable, handheld scanner and detachable palette, which is easy to use and takes only a small fraction of time to measure the leaf area, directly using leaf.

The plant canopy imager or image scanner with image analyzing software captures wide-angle canopy images while estimating Leaf Area Index (LAI) and measuring Photosynthetically Active Radiation (PAR) levels. In daylight condition the self-leveling digital camera takes 150° images of plant canopies and the 24 photosynthetically active radiation (PAR). Sensors in the wand of the device measure light to calculate Sunflecks. The handheld tablet computer powers the wand and displays a live-updating image and plant canopy calculations simultaneously.

Max-min thermometer and hygrometer   Light meter

pH & moisture meter   Plant canopy imager

4

Design and Construction of Greenhouse

Design and Construction of a greenhouse involves five distinctly different aspects viz. (a) Basic form or type, (b) structure/designs, (c) covering materials and its fixing, (d) foundation and constructional detail, and (e) crop-support, irrigation and drainage planning.

All the above aspects are interlinked. To have a clear idea, these individual aspects are discussed separately.

4.1. Basic Forms of Greenhouse

Before considering anything about construction, one should finalize the form of construction. There are different forms of greenhouses for different climatic and crop situations. These types of structures are planned according to the requirement of farmer/promoter and obviously according to the purpose. Now we will describe, in brief, some of the important form or type of greenhouses.

4.1.1. Greenhouse covered with transparent and impervious material

These structures are covered with water/rain impermeable transparent materials which allow necessary solar radiation for optimal growth of plants. Simultaneously it protects crops from temperature-fluctuation, rain, wind, and creates a favourable situation in respect of CO_2 and humidity. This covering is also used to prevent the re-radiation of heat energy and conserve the heat generated in daytime to raise night temperature inside.

The roofs of such structures always provide sufficient slope to allow the precipitation, in all forms, to run off safely out of the greenhouse area.

Different types of impermeable and transparent materials can be used to cover a greenhouse. They may be rigid or flexible. However, now, only the flexible transparent plastic films are used commercially as covering material due to its cost effectiveness. So we will focus our discussion on covering materials of flexible plastic materials only.

Most popularly used material to cover greenhouse is thick polyethylene (PE) UV treated films. These types of greenhouses are generally called as poly-houses. (The detail about these materials is discussed in the chapter-4.3.1)

Based on different climatic situation different forms of poly-greenhouses are planned, which differs in respect of area and type of opening/ventilation.

a) **Naturally ventilated Poly-house or poly-greenhouse with closable small ventilators:** These greenhouses have ventilators/openings at the top or ridge of the roof to vent out hot air from the greenhouse. Sometimes the opposite side walls are also provided with ventilators to offer cross ventilation. All these ventilators may have closing arrangements to conserve inside high-temperature to solve the problem of low temperature during night.

These types of structures are generally constructed in climatic areas of sub-tropical highlands and dry areas where extreme minimum temperature frequently sinks below 0°C at night. On the other hand most of the commercial nurseries also use such type of greenhouses, irrespective of climatic condition, with proper climatic control system and provision of monitoring the inside climate as per requirement.

Some of these forms (high-technology greenhouses) are of complicated type where controlling of temperature, humidity, CO_2 status of air is managed with high precision. The closing of ventilators are mostly done by mechanical means and controlled automatically through electrical or computerized devices.

These structures are costlier and should have heating or cooling or both arrangements. The covering material may be thick or double layered to protect the heat loss through re-radiation in cool climate.

The percentage of ventilation area against floor area of such structure will be around 20% or less. However, according to the crop-climate requirement the actual ratio is to be determined.

Low profile or low height (3m) structures are used for such types of greenhouses.

b) **Naturally ventilated Poly-house with large ventilators with roll-up curtains:** In dry tropical and dry-hot sub-tropical regions, where average night temperature is not below 10° C and the extreme low temperature does not go below 0° C the ventilation of greenhouse may be done through larger openings at ridge of the roof, sidewalls and gables. The roll-up curtains are used to close the openings whenever necessary. The closing mechanisms can be done by mechanical means or by power or can be operated manually.

These openings may be covered with insect proof net. No heating mechanism is required for this type of structure. The efficient natural ventilation controls the temperature, humidity etc to a large extent. If the natural ventilation is not sufficient, exhaust fans and evaporative cooling systems like fan-pad cooling and misting/fogging are used to bring down high inside temperature to the desired level during very hot period.

In this type of greenhouse the ratio of ventilation area to floor area is around 0.5. These structures are generally 'taller and in an average 4 to 4.5 m gutter height is recommended.

c) Naturally ventilated Poly-house with permanently open large ventilators: In dry tropical and humid/wet subtropical climates, where both average night and average winter temperatures are not much low (10^0 to $15°C$) and the extreme low temperature at winter night does not drop below $5°C$, these structures of greenhouse works better, particularly in the wet sub tropics where average rainfall is very high. These structures immensely help crop production in rainy season. If manageable, insect proof nets may be used to cover the openings. Sometimes the entire opposite sidewalls may be kept open to create a tunnel effect inside the greenhouse.

No heating or cooling mechanism is required for such structures. Only the shading arrangement (with shade net) is needed. This removable shade helps to cut down the high intensity solar radiation in daytime reducing the temperature to a good extent.

It is also a taller structure with necessary roof vent at the ridges and the ratio of ventilator to floor area will be 0.6 or more. Single structure of more than 1000sqm size may not be recommended to avoid proper natural ventilation process. However, in case for larger size greenhouse a central open space may be provided.

d) Retractable roof greenhouse: It is a new idea of greenhouse designing and construction, where the roof cover can be removed whenever necessary. It can be partially removed as per requirement. It is basically a multitask greenhouse and can perform as different types of greenhouses as the situation requires.

It is very much suitable for hot climate and can be designed for both dry-hot and humid-hot climates. This structure can also act as both form of greenhouse simultaneously, as a poly house and as shade house. Furthermore it can also act as an open field, if the situation demands, like inviting pollinating agents at the time of release of pollen from pollen sac etc.

e) Shelter-structure poly-house: The smaller or series of small structures of poly-houses having slanting polythene roof with permanently open sidewalls are called shelter poly-house. The roof is constructed to protect the crop from rainfall only from the top.

In the tropical area close to equatorial zone (0° latitude), where the difference of average day-night and summer-winter temperature is minimum, and average solar-radiation and rainfall is high, this type of shelter-structure poly-house is recommended. These structures may be small in size and are suitable in hilly terraces. Such structures are useful in Central Africa, northern parts of South America, parts of Southeast Asia etc.

These structures protect the crop from impact of high intensity rain and can also protect crops from high intensity solar radiation if shading-net is used.

4.1.2. Greenhouse covered with porous material (Shade/net-house)

These greenhouses are covered with shade providing materials, which are water/rain permeable. These structures protect the crops from strong solar radiation and heavy impact of rain and wind and also from high and low temperatures.

Like permanently open ventilator poly-houses, these are suitable for hot tropical and subtropical areas.

In general the roof of such structures is flat or horizontal. However, curved roof or Quonset type structures can be designed where polyethylene or mono-filament shade-net is used as covering material. Commonly the woven shade-nets made from polyethylene tape/raffia are used as covering material for such shade-house. (The details are given in chapter- 4.3.2.).

BOX- 8

Polyethylene- Polyethylene (PE) is a thermoplastic polymer with variable crystalline structure and an extremely large range of applications depending on the particular type. It is one of the most widely produced plastics in the world. The commercial process (the Ziegler-Natta catalysts) that made PE such a success was developed in the 1950s by German and Italian scientists Karl Ziegler and Giulio Natta.

It is made from the polymerization of ethylene. Ethylene (C_2H_4) is a gaseous hydrocarbon commonly produced by the cracking of ethane, which in turn is a major constituent of natural gas or can be distilled from petroleum. Ethylene molecules are essentially composed of two methylene units (CH_2) linked together by a double bond between the carbon atoms—a structure represented by the formula $CH_2=CH_2$. Under the influence of polymerization catalysts, the double bond can be broken and the resultant extra single bond used to link to a carbon atom in another ethylene molecule. This simple structure, repeated thousands of times in a single molecule, is the key to the properties of polyethylene.

The long, chainlike molecules, in which hydrogen atoms are connected to a carbon backbone, can be produced in linear or branched forms. Branched versions are known as low-density polyethylene (LDPE) or linear low-density polyethylene (LLDPE); linear versions are known as high-density polyethylene (HDPE) and ultrahigh-molecular-weight polyethylene (UHMWPE).

The LDPE is used for manufacturing of greenhouse films and HDPE tapes are used to weave agro-shade-net.

Monofilament— Monofilament yarn shall be manufactured from HDPE granules (see IS 7328), which shall be UV stabilized by adding suitable UV stabilizer. The linear density of the monofilament yarn shall be 33.33 tex (300 denier) for Type I and Type II shade nets and 55.55 tex (500 denier) for Type III and Type IV shade nets. As agreed to between the buyer and the seller, coloured monofilament yarn shall be manufactured using colour master batch. The denier of HDPE monofilament yarn used in the manufacture of shade nets shall be subjected to the following tolerances: a) 10 percent on individual value; and b) 5 percent on average. The heat shrinkage of the monofilament yarn at 60°C shall not exceed 5 percent and shall not exceed 8 percent at 95°C. For determining the shrinkage, the monofilament yarn shall be subjected to the specified temperature for a period of 10 min in an air circulating oven and hot water bath respectively. The fabric used in the manufacture of mono filament yarn shade nets shall be knitted on the raschel knitting machines and shall have a width as per the agreement between buyer and seller. A tape of 1.70 ± 0.1 mm width or monofilament yarn of suitable diameter having the name of the manufacture and applicable shading factor shall be provided at one end of the fabric for proper identification of manufacturer.

Two basic types of covering materials are used for shade-houses: (1) dry plant materials and (2) Polyethylene nets.

For few specific crops, in few specific areas, such structures are still constructed with dry plant materials. For example a shade-house called 'Boroj' for production of betel-leaf is still made out of dry plant materials in south East Asia. Otherwise use of polyethylene net is the only commercial material used for construction of a shade-house.

However, these structure or shade-houses may be divided into two types:

(i) **Fully covered or hall type shade-house:** Hot and dry climate where average night temperature is moderately low (below 15°C), such structures are recommended. These types of structures are covered by shade-net from all the sides permanently. 'Boroj' is also this type of shade-house. Apart from protecting crops from strong solar radiation and impact of heavy rain, this structure also protects the plants from hot wind, birds and larger insects. By way of reduction of solar radiation and permitting a portion of re-radiation back to atmosphere, the balance of day-night temperature is maintained in such greenhouses. However, spraying or misting of water can be done in hot dry mid-day period to bring down high temperature if necessary.

Such structures do not face any kind of problem related to CO_2 deficiency at daytime. On the other hand, the shade provided to the crop also controls the excess evapo-transpiration saving the crop from facing temporary wilting due to excess transpiration.

To grow shade-loving plant, shade-house is an essential structure.

(ii) **Partially or temporarily covered shade-house**: These structures are recommended in the hot humid areas where difference of day-night temperature is not very much and average night temperature is not below 15°C. It is a very simple structure where the sidewalls are not covered and remain open or have a removable covering with shade or insect-proof net. Here only the roof will be permanently covered with shade net.

Sometimes, particularly in cloudy weather, the roof cover (shade-net) can be removed temporarily. If necessary, plastic film may be placed to protect the crop from heavy impact of rain along with the shade-net. In such **hybrid structures** the roof may be ridge & furrow or saw-tooth type having necessary gutter in between. It may or may not have necessary ventilation at the top. Both the shade-net and plastic film fixed at the top will be removable and placed according to the requirement. Such structure with high-end technology is called 'Retractable greenhouse', which is expensive also.

4.1.3. Other forms of greenhouse

Before going into detail, we should discuss some points that are very much essential to finalize the structural design of a greenhouse.

It is very difficult to give a specific set of requirements to prepare the structural design of a greenhouse, because there are many exceptions to any rule due to different unpredictable reasons, particularly when climate is associated with. Still the following points should be considered while making a structural design of a greenhouse.

- The structure must meet the engineering norms of the specific locality.
- The specification of constructional design shall be done by an engineering firm or an engineer according to the proposed requirement.
- The large greenhouses are usually custom designed and built accordingly.

However, it is important to understand the basic engineering design of the greenhouse. It is different from the basic principal of civil engineering, in respect of load calculation necessary for construction of house etc.

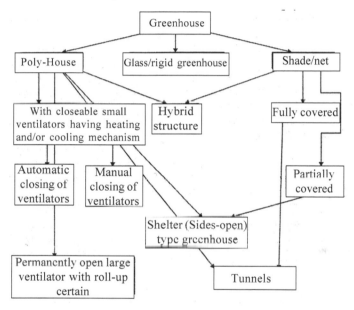

Fig. 12: Types of greenhouse on the basis of covering material

- The structural design should provide maximum light transmittance i.e. minimum obstruction to light entry, through roof and wall, while considering support members to the roof and structure.
- Calculation of design load - the design load includes the dead load and the live load. The dead-load includes the weight of structure, glazing/covering material, permanent equipment and machinery attached to the structure etc. The live-load includes the weight of people working on roof, hanging plant and shading materials, precipitation (rain/snow) load and the most important wind loads. Out of all these loads wind-load is the major element to be considered for greenhouse construction, which otherwise is not at all important for house construction up to 10 m height.
- The gutter should slope slightly to encourage drainage of roof run off and the gutter height should be enough to perform all operation freely in gutter-connected greenhouse.

- The drainage system of surrounding areas along with the drain for carrying the roof runoff should have been properly designed. It should be designed in such a way that the precipitated water should not enter into the bed or root media through seepage or by any means.
- The irrigation system in particular and the operational work should be compatible with the width of the greenhouse bays. Accordingly the walkways or driveways are to be designed, which will allow free movement spaces.
- Proper calculation for the foundation of greenhouse and transfer of loads to the poles is very important.

However, the constructional design of a greenhouse consists of two independent parts i.e. the load bearing foundation and the design of roof and sidewall. The constructional design of foundation, roof and sidewall is discussed separately in chapter-4.4.3. This part of the chapter, will deal with different kind of structural types of greenhouse.

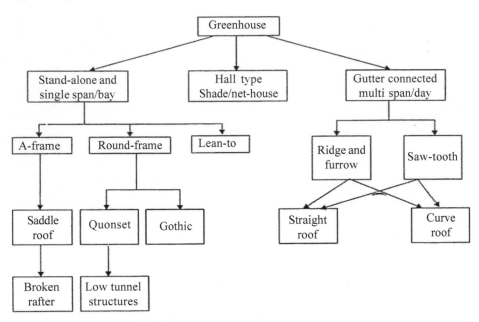

Fig. 13: Different types of greenhouse on the basis of design

Various types of greenhouse structures have been practiced and tried throughout the world. Earlier most of them were designed for extreme climates, initially for snow falling areas and then for extreme hot climates. Now the other areas, particularly the tropical and sub-tropical areas, are gradually getting more importance in respect of designing the structure of greenhouse suitable for that area.

However, considering the gulf changes occurred in greenhouse technology after introduction of PE film and extension of utilization area, the basic structural types that may be useful today are discussed below.

4.2. Structural Types of Greenhouse

4.2.1. Stand alone or single span/bay greenhouse

It is basically a structure that covers a small piece of land usually, but not always, having no pole or post inside. Depending upon the roof, these structures are generally of two types: 'A' frame type and 'round' frame type with necessary variations. These structures may be modified as per need or purpose and sometime to accommodate climate control mechanisms or more space inside. In early days when glasses were used as covering material, A-frame structure was commonly used. After introduction of PE-film the round-frames are gaining popularity. Now we will discuss the detailed structural design of the different stand-alone greenhouses.

(a) Saddle-roof or even-span: A-Frame structures: these types of structures are more suitable in cooler climates. The simplest saddle roof A-frame structures are made up of post and rafter only. This type of construction consists of an embedded post connected with *rafters*. It requires more structural materials/ elements than some other designs (round type). Strong sidewall posts, deeply embedded, are required to withstand outward forces of rafter and wind pressure. These types of structures provide more space and effective ventilation (Fig. 14).

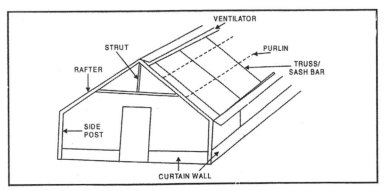

Fig. 14: Saddle roof, 'A' type greenhouse and its different structural components.

There is another type of structure, very much similar to the post and rafter type except that it has a collar beam, which ties the upper parts of the rafters together to form a truss like structures. The *purline* or horizontal numbers are usually connected to these series of supporting *trusses* to form the *gable* of a roof.

Sometimes to construct a stronger roof the tie beams (which is excess for common poly-house structure) are connected both to the rafters and the small vertical beam placed in the middle of the tie beam (kingpost). Sometime few other supporting vertical members are used to form the stronger trusses useful for wider greenhouses having heavy covering material like glass. The strength of this structure primarily comes from the trusses set on the vertical posts of the side wall. The weight of the structure along with the covering material of the roof and other stresses/loads are borne by the trusses and transferred to the side walls that in turn transmit the stresses to the ground.

The spacing and placement of rafters and collar beams, which form the trusses, depends upon the capacity to withstand the pressure of precipitation and wind received by the covering material. A plastic film covered A-frame stand alone type greenhouse, made of wood, was designed (as early as 1950s by University of Ketucky) in early days which have changed its design with years. However, it is still very much relevant, particularly when a new design is created using materials like Bamboo. Thus one should know the basic design of that old structure.

Width	- 20 to 30 feet, maximum up to 40 ft.
Side posts and columns	- 4" x 4" lamber, No internal post.
Rafters	-2" x 3" or 2" x 4" alternated with 1" x 4" members
Spacing	-30" to 36"

If the wind load to the greenhouse is higher, then a support (in the form of angular pillar) may be provided to the side posts, particularly to the corner posts.

In general these types of structures are typical with *skirting wall* or side guard wall made up of concrete, cement, brick etc. up to the height of the crop i.e. generally up to 2'4" height. Now a days such wall is created with PE-film and called as '*Apron*'. This wall or apron somewhat reduces the air movement beneath the crop canopy.

Roof vent is sometimes provided to vent out the hot air in high radiation period. These vents are closable and are kept open whenever necessary.

(b) Broken-roof/span A-frame structure: These structures are suitable for warmer climate. The principal difference of this structure from the saddle roof or even span type is that, it has uneven and unequal roof span or rafters. Sometimes one span may be missing completely to provide the basics of saw-tooth type roof.

For construction of these structures, the post and rafter system are used in an improvised manner to modify the design of roof. Here the two rafters of a truss are not attached at the top as it is fixed in the saddle roof structure. One rafter is projected upwards and the other converges with the king-post 1 to 3 ft below the top of the roof where the other rafter is fixed (Fig. 14a). The truss of the roof is constructed with the king-post set on the tie/collar beam and two unevenly projected rafters. Thus construction of uneven span roof-structure is completed. Otherwise, the rafter with smaller angle is connected with the other rafter at 1 to 3 ft below the middle-top point of the roof (Fig.14b). This also gives the same design created using king-post, as mentioned earlier. The vertical drop area of the two roofs generally remains open and used for ventilation purposes, which is called 'roof window'. It helps to create the 'chimney effect' necessary for greenhouse.

The load bearing capacity of these roof structures, which is somewhat less than that of saddle -roof structure, is dependent on the strength of tie/collar, beam. However, for polyethylene covering material this structure works quite well. To provide additional strength to this type of truss, a pillar may be added below the kingpost to provide additional support to the truss. This is done either to provide more width to the greenhouse or to reduce the amount of wood or metal required for roof structure.

If the locality have a higher pick wind-load then an angular support to the side posts, particularly in corner posts, are essential, sometimes it is called Hockcy.

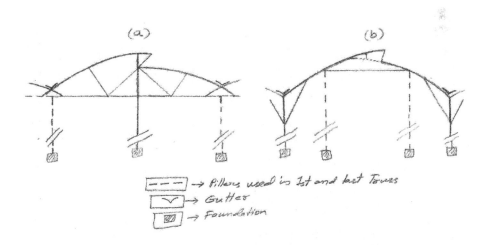

Fig. 14 a &b: Sketch of two different structural design for roof ventilation

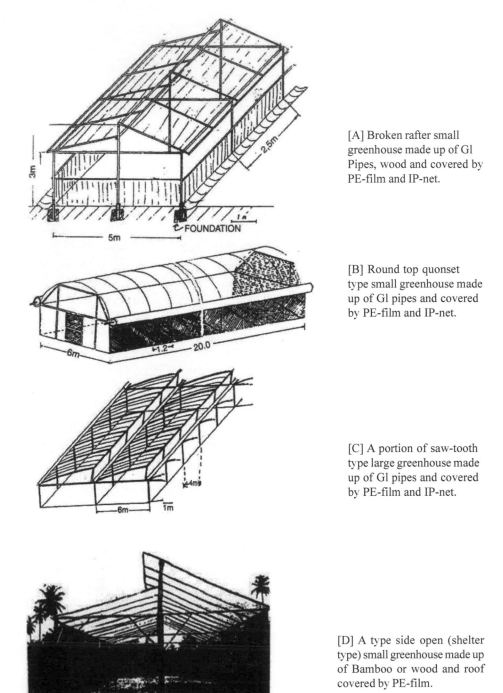

[A] Broken rafter small greenhouse made up of Gl Pipes, wood and covered by PE-film and IP-net.

[B] Round top quonset type small greenhouse made up of Gl pipes and covered by PE-film and IP-net.

[C] A portion of saw-tooth type large greenhouse made up of Gl pipes and covered by PE-film and IP-net.

[D] A type side open (shelter type) small greenhouse made up of Bamboo or wood and roof covered by PE-film.

Fig. 15: Few models of low-cost Natural Ventilated Greenhouse suitable for tropical humid climate.

Sometimes one span of this kind of structure is missing completely. This forms a slanting single span roof structure that somewhat resembles the greenhouse attached to a wall of a house or garage, generally known as lean to type greenhouse (Fig. 16).

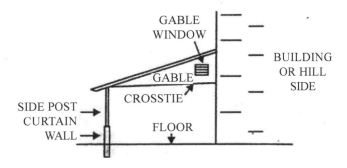

Fig. 16: Construction of a typical lean-to greenhouse

(c) Quonset or hoop type round frame structure: This structure is suitable for cool climate, where there is no snowfall even in the coolest period i.e. in winter. It is a roundish or 'D' shaped structure based upon an arched roof. The arched roof allows stresses on the structure to be efficiently transferred down to the ground. In general the arch frame of this structure extends up to the ground with no sidewalls. Sometime a small guard wall is provided along the periphery of the greenhouse.

The construction procedure is easy and inexpensive in relation to the A- frame structures. Simple and efficient structure made up with galvanized steel pipes or bamboo (in case of low cost) frame maintaining a middle-height of 10ft to 14ft is generally recommended. The pipes or bamboo frames are bent according to the necessity to form round structural frame. In general, the pipes are bent to fit a 180° arc modified for somewhat more vertical sides. The ends of the pipes are fixed firmly on the ground and a 2inch x 8inch wooden plank or galvanized bar is attached to the base of the pipes along the length of the greenhouse keeping no space between runner and the ground surface, virtually the wooden plank is partially buried into the ground. Otherwise the covering material (film etc) of apron is buried in to the soil along the side wall of the greenhouse.

The pipe arches or trusses are supported by purlins, made up of pipes or reinforcing wire or suitable materials, running through the length of the greenhouse. The arches are generally 3' to 4' apart from each other. The covering material, generally the PE-films (single or double layered), is then put and fixed on the structure to form the greenhouse. Agro-shade net is also used as covering material for these types of structures.

Sometimes for bamboo structure or for big size pipe structure with an intention to reduce quantity of metal, a middle line post is used. However, sometime the Quonset type of structure does not allow proper head room in the sides of the greenhouse, which restrict cultivation and other operations along the sidewalls of the greenhouse.

To solve the problem of headroom along the sidewall an improvised structure of Quonset greenhouse is used. In such structure the arch essentially forms roof and gable section, which is directly set on the straight posts of the sidewalls (Fig. 14). The design of sidewall of Quonset type shall provided suitably by using angular inside support and may be done in the line of A-frame structure.

(d) Gothic type structure: It is similar to Quonset structure but has a gothic shape i.e. distinctly pointed at the top of the arch instead of being round (Fig. 17).

This makes the house taller having more slanting roofs on both the sides of the pointed top of the ridge. This will help the condensed water to flow better through the inner side of the covering restricting water drops to fall over the crops. The precipitation also flows quicker through the roofs. This house provides more headroom along the sidewalls. Frequently, the pointed arch roof is set on straight vertical posts of the sidewalls.

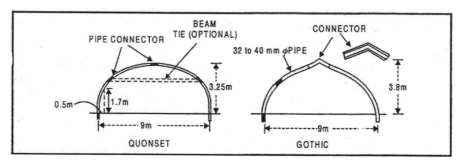

Fig. 17. Two models of large Quonset and Gothic structure of greenhouse

4.2.2. Gutter connected or multi-span/bay greenhouse

This is a series of stand-alone or single-bay greenhouses connected or joined at sidewalls through a gutter without having a partition wall. This brings most of the area under cultivation and act as a large greenhouse having unobstructed interior under a single stretch of roof.

Most of the modern commercial greenhouses use this type of structure or some variation of such structure. The scope of space utilization and mechanization of different operational activities is more than any stand-alone structure.

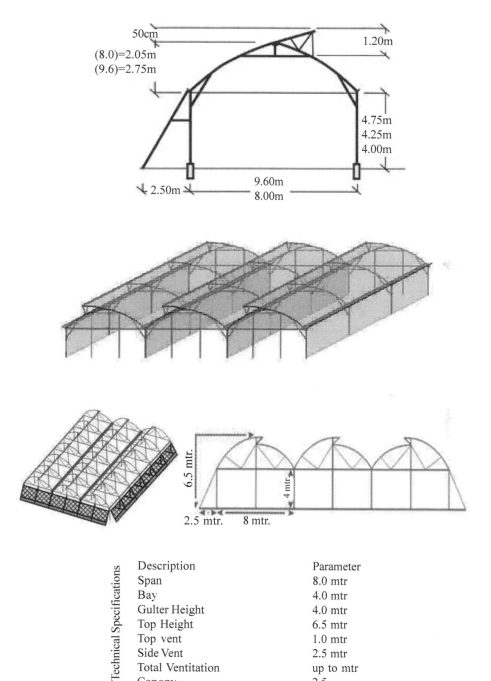

	Description	Parameter
	Span	8.0 mtr
	Bay	4.0 mtr
	Gulter Height	4.0 mtr
Technical Specifications	Top Height	6.5 mtr
	Top vent	1.0 mtr
	Side Vent	2.5 mtr
	Total Ventitation	up to mtr
	Canopy	2.5

Fig. 18: Basic structure and dimension of commonly used large naturally
ventilated commercial greenhouse

However, there are few problems associated with gutter-connected structure. For example this structure is not much suitable to maintain different environmental conditions required for different crops. That means cultivation of different types of crops under one gutter connected greenhouse is difficult. At the same time to maintain a uniform inside climate through naturally-ventilated or other climate control mechanism is also a problem in such type of greenhouses. To solve this problem rollup curtains can be used to increase the efficiency of cooling or heating mechanisms.

Two types of gutter-connected greenhouses are used in different parts of the world, viz. high-profile and low-profile.

The high-profile or taller structures having larger roof and side-wall windows are used generally in warmer climates, which efficiently vent out the hot air through roof window and provide entry of more fresh air through side windows. As a result inside air-circulation is increased that helps to reduce the inside temperature.

The low-profile, i.e. relatively shorter greenhouses having smaller roof and side-wall windows, are generally used in cooler climate. Such design is useful to conserve heat radiation to increase inside temperature.

Depending upon the design of roof structure, now-a-days, two distinctly different gutter-connected commercial greenhouses are constructed, which are (1) ridge and furrow structure and (2) saw-tooth structure.

(a) The ridge and furrow gutter-connected greenhouses: These types are constructed by joining series of single even/uneven span stand-alone structures. The name of the structure suggests the feature of the roof where the gutters act as furrows and the raised middle portion of roofs made up of even/uneven spans of roof constitute the ridges. Many variations of this structure are available and practiced in different parts of the world. However, uneven span ridge and furrow gutter-connected greenhouse is most common that can provide effective roof window.

The top level of roof spans of such structures may be straight or may be curved (Fig.19 a & b). For wider span and to increase load bearing efficiency the curved roofs are more useful.

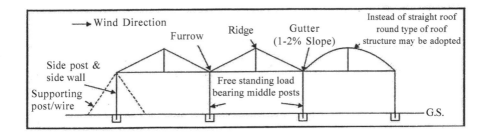

Fig. 19(a): Ridge & furrow type greenhouse, with saddle or round roof.

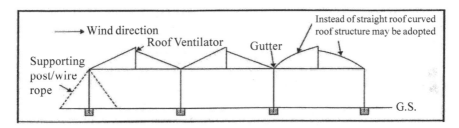

Fig. 19 (b): Ridge & furrow type greenhouse with 'A' type or round type broken rafter roof structure

(b) Saw-tooth gutter connected greenhouses: This type of gutter-connected structures has recently gained popularity particularly in warmer climate where more efficient roof ventilation is necessary. Here a single bay is covered by only one straight or curved span of slanting roof i.e. the second span of roof in each bay is missing completely. The gutters are placed at the lowest point of the roof span (Fig. 20 a & b).

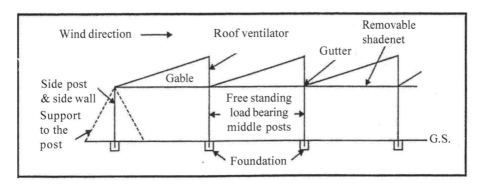

Fig. 20(a): Straight roof saw-tooth greenhouse

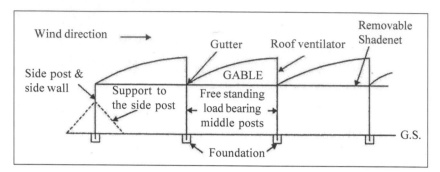

Fig. 20(b): Curved roof saw-tooth type greenhouse.

4.2.3. Tunnel Type Greenhouse (for detail see heading 4.6)

It is a low height completely closed 'half-round'-shaped or Quonset type structure. These types of structures are frequently used for raising seedling, rearing of plants, and also for growing summer crops in winter season. The covering materials (PE-film or shade/IP-net) of this structure may be for single use or designed as removable whenever necessary and have no permanent ventilation. Two types of tunnel type greenhouses are commonly used, which are 'low-plastic tunnel' and 'walk-in tunnel'.

The 'low-plastic tunnel' may be defined as "the low height temporary structure covered by any thin transparent material (50-100μ PE-film, mostly for single use) or insect-net. The structure is made to envelope only one row or bed, up to desired length, to enhance the plant growth under it". [Fig. 21 (a)]

Primarily transparent plastic films are used for this structure, which helps to increases temperature around plant, thus helping the summer crop to grow better in winter. It also advances the maturity of crop by 30 to 45 days. However, where the temperature difference between summer and winter is not so wide and where the average night temperature in winter season does not drop below 10°C, this structure does not yield the expected result.

Sometimes the low-plastic tunnels are used to save the young seedlings from heavy splash of rain and from insect-pest and disease infestation. Generally these tunnels are so small that for any cultural operation removal of the covering is required. In most cases no intermediary cultural operation is carried out during the life span (1 to 3 months) of these structures. The heights of such structures are 2 to 5 feet with a width of 2 to 7 feet.

On the other hand, walk-in tunnel greenhouses are permanent type low-height (man-height) structures, where both ends of the tunnel may be provided with windows and are utilized to increase air circulation / flow inside [Fig. 21(b)]. This structure has a good capacity to create 'tunnel effect' inside when the both ends are provided with windows or remain open and the orientation is in the direction of wind.

Greenhouse for spacial purpose

500 sqm aero dynamic greenhouse

Saw-tooth type greenhouse

Ridge and furrow type greenhouse

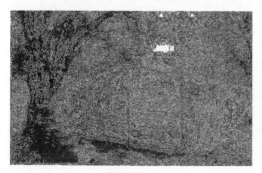

Shade house   Insect-proof net-house

Application of shade-net over roof of the greenhouse

Outside & inside view Low-plastic tunnel

White colour Shade-net house

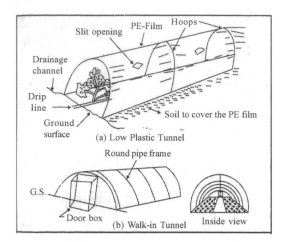

Fig. 21: Model of (a) low-plastic tunnel and (b) walk-in in tunnel

Urban or Hobby type greenhouse

Using any of the common structural and covering material and following the basic principle of design, as mentioned above small or mini greenhouses can be made for small urban family to cultivate desired crops for their own use. The crop may be vegetables or fruits for own consumption or for ornamental plants or flowers for hobby. Wood, bamboo, steel, or any structural material can be used for such greenhouses. If such greenhouse is to be placed on the house roof, the adequate measure should be taken to protect it against strong wind or storm. For tropical climate the efficient naturally ventilated system should be incorporated in the design of such urban or mini greenhouses (see fig.22 & 23)

Such greenhouses may be provided with necessary beds or pots or vertical farming system, as shown below for the plants and the irrigation or fertigation system. Such greenhouses are sometimes sufficient enough to supply fresh and perishable food items like vegetables etc to the kitchen directly for the family.

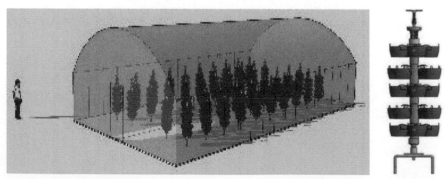

Fig. 22: Model of vertical farming (pole) system in urban greenhouse

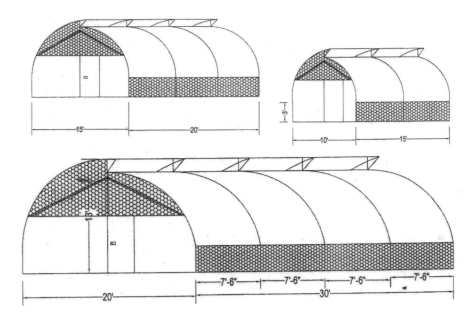

Fig. 23: Model Urban greenhouse of different size

4.2.4. Shade/net greenhouse

It is the simplest permanent hall type structure of greenhouse (Fig. 32). As the covering material is porous, there is no need to build a strong and slanting roof structure like poly/glass- greenhouse. This type of structures is made up of pillars, spaced 4 to 5 m apart, and wire rope or suitable structural materials placed at the top/roof. They are used to connect the pillars that finally carry the covering material i.e. shade-net or insect-net. Thus, depending on the type of covering material such greenhouses are termed as 'shade-house' and 'net-house'.

4.2.5. Advantages of the four basic types of greenhouse

(1) Stand-alone/single bay greenhouse

1. Usually best for small grower, planning on less than 1000 square meter of floor area. If business grows, additional house may be constructed. Thus, expansion as per requirement is easier.

2. Easier to use for specific purpose and install specific climatic control mechanisms in each greenhouse as per requirement. The naturally ventilated greenhouse system is more efficient in smaller structures.

3. Best suited for heavy snow area and also for undulating/hilly area.

4. Crop control, particularly the plant protection measures are easier to handle.

5. Easier to build and maintain. Less expensive particularly when local material for structure is used.

6. Shifting from one purpose to another is easier.

(2) Gutter-connected or multi-bay greenhouse

1. Useful for greenhouses with more than 1000 sqm of floor area.

2. More cost effective.

3. Heating cost is less due to less surface area to floor area ratio that reduces the transmission of heat from the crop of the greenhouse.

4. Provide more effective growing area

5. Greater labour efficiency can be achieved in such structures.

6. Easier to install and automate the electrical service, water supply/irrigation system. The operation cost will also be less.

7. Easier to automate the environmental control mechanisms.

8. Proper ventilation reduces the cooling cost particularly in high profile structures.

(3) Shade-house and Net-house

1. Shade-house is useful in high solar radiation areas and net-house is useful in humid areas where the chance of viral-disease and pest-infestation is high.

2. Easiest and cheapest to construct.

3. With least environment control mechanisms it can create a microclimate inside in favour of the crop.

4. Shade-house is very much useful for shade-loving crops.

(4) Low plastic tunnel and walk-in tunnel

1. Selected crops can be grown in off-season or in advance (in winter with very low temperature) without much extra effort and little more expenditure than that of open field condition.

2. Both the low-tunnel and walk-in tunnel are utilized efficiently for raising disease free healthy seedlings.

3. This structure raises soil temperature and temperature around crop that enhance growth and yield of the crop advancing or extending the crop life by 30-40 days in comparison to open field.

4. Generally, in low-plastic tunnel, cucurbits like melons, gourds, squash etc are grown in winter that provides a very good return.

5. Thin plastic film and horse-shoe shaped hoops (made up of flexible galvanized or plastic coated/covered iron, or bamboo) of desired size

(generally 2-3ft height and 3ft wide), spaced 1.5 to 2m apart, are the only materials required to construct a low- plastic-tunnel.

6. To build a walk-in tunnel, almost all the materials, required for normal greenhouse, are used. So it will be expensive than that of low-tunnel and cheaper than that of normal greenhouse.

4.3. Covering Materials and its Fixing

Types of covering materials

The discussion about the planning of greenhouse categorically divides the greenhouse covering materials into two groups.

1. Transparent and water impermeable materials;
2. Non-transparent and water permeable materials.

Now we will describe in details about these two groups of covering materials separately.

4.3.1. Transparent and water impervious materials

Different types of impermeable and transparent materials can be used to cover a greenhouse. They may be rigid or flexible.

Initially glass was used to built greenhouses. Then, after use of plastic started, different kinds of rigid plastic materials were also used to cover greenhouse.

After introduction of flexible transparent plastic films the use of glass is being gradually discontinued for commercial purposes due to its high cost. However, for experimental and other specific purposes the rigid plastics like 'polycarbonate sheet' etc is used to cover greenhouse.

However, we mainly concentrate our discussion on flexible transparent plastic films.

There are two sub-groups of these transparent covering materials:

(i)   Rigid transparent covering material;

(ii)   Flexible transparent covering material;

(i) **Rigid transparent covering material:** Glass, polyvinyl chloride (PVC), Fiberglass reinforced plastic (FRP), Acrylic and polycarbonate rigid panels are used in a few commercial greenhouses particularly in cool climates. These materials are used in the form of panels of different sizes and shape (Hat, corrugated) and are attached to the frames along the slope of the roof called sash bars. The different rigid materials have different characters like longer life, low heat loss etc. but the initial high cost and maintenance make them uneconomic as well as unpopular, specifically in tropical areas. Most importantly in comparison with flexible films of plastic (principally polyethylene), this rigid material does not have any acceptability to the people due to too high initial cost. So we will not discuss about the rigid covering material in details.

(ii) **Flexible transparent covering materials:** The most preferred covering material for greenhouse nowadays is the flexible transparent plastics. Out of different flexible plastic films like Mylar, Vinyl, Polyethylene (PE), the PE-film is principally used for greenhouse coverings. It is of two types, clear and opaque (diffuse).

There are several advantages of PE over different rigid covering materials including glass.

- Low cost (for saw tooth structure about 1 kg PE-film is required for covering about 5 sqm area of greenhouse);
- Low structural expenditure due to less supporting frame work;
- In case of two layer covering, it provides better insulation;
- Easy to handle due to light weight;
- Minimum joint in covering provide better scope to stop leakage;
- It passes all wavelengths of light required for the growth of the plants;
- Polyethylene film permits the passage of oxygen and carbon dioxide while reducing the passage of water vapour.

Problem and Solutions

Durability: The only problem with this PE covering is its durability. The thinner (0.1 mm) film is required to be changed every year and the thicker (0.2 mm) may be changed every alternate year. However, the ultra-violet (UV) inhibitor 0.2 mm PE films can last three to four years.

Heat loss or transmissivity of heat: The single PE film covered greenhouses lose more heat at night or in winter in comparison to rigid covering materials. So in cooler climates two layers of PE-film is used to cover greenhouse, which retain more heat (the heating bill will reduce up to 50%) than the glass greenhouse.

BOX - 9

Double layer covering of greenhouse

The two layer of UV-stabilized PE-film with a 2.5 cm (1 inch) air gap between layers is kept separated by injected air pressure. A small blower will do the job perfectly. However, larger air gaps, up to 10 cm (4 inch), separated by frame of roof structure may serve the purpose, but with higher cost and less efficiency. Below 2.5 cm (1inch) air gap diminishes the insulating property and below 3/4th inch gap actually have no effect, as because when the two layers touches each other, the insulating property is lost. The outer PF-film layer of such two-layered coverings may be thicker as because the inner layer receives sunlight of low UV level.

In general these PE films are available in tube form, which is cut open through length to use as a single layer covering. Widths of such tubes are 2.5 or 3 or 3.5 m, which covers 5 or 6 or 7m span of gable of a greenhouse respectively.

In tropical wet and dry and sub-tropical wet areas single layer covering is suitable and sufficient. The greenhouses of these areas, where ventilators are open throughout the day and night, the factor of transmissibility of long-wave radiation (IR) from the greenhouse are not of serious concern. So single layer UV-stabilized PF-film may only be used in such areas.

BOX - 10

Transmissivity of long-wave radiation (IR)

It is the capacity or quality of the covering materials to transmit the long wave (more than 3000 nanometer) re-radiation emitted by plants, soils and structural materials of greenhouse. This Long-wave radiation (IR) has an influence on the plant temperature and on the heat transfer through the roofing material. The highest radiation intensity of energies emitted by the plants, soil & installation in the greenhouse lies in a wavelength range between 600 and 14000 nanometer.

The films having high IR-transmissivity allow more heat loss from the greenhouse, thus lowering down the plant temperature in comparison to film with lower IR-transmissivity. In this situation, at cool nights with clear sky, the plant and air temperature inside can fall below the outside temperature, which can damage the crop.

Light transmissivity of PE films: It is generally referred in percentage of outside sunlight. Initially all PE-films of 0.1mm to 0.2mm thickness have the transmissivity between 89 to 92 percent PAR (photosynthetically active radiation) light. However, in case of UV stabilized PE-films it sinks by 5 to 7 percent in two years.

In case of two layer covering with one clear film (have 90% PAR) and one layer of IR (thermal)-film (have 87% PAR), 78.3% of PAR light should be realised.

Types of PE films

1. Non stabilized PE-films: This is a non-stabilized form with maximum durability of two years. Its thickness ranges from 100 micron(μ) (0.1mm) to 200μ (0.2mm) of which 0.2mm thick films may last maximum up to two years depending on the climate. However, these films are not recommended for hot tropical regions of high intensity solar radiation. Seasonal vegetables or flowers in wet tropical and subtropical areas may be cultivated under such covering materials for one year only. If this is established as a financially viable option, like for low-tunnel, then it is recommendable to change the film every year. Due to high intensity solar radiation these common UV un-stabilized PE-films become brittle more quickly and its light transmissivity is reduced and naturally becomes susceptible to wind.

2. UV-stabilized PE-film: This film does not allow the UV radiation (up to 381 nanometers where 400 nanometers marks the end of U V radiation and the beginning of visible light) to enter the greenhouse. This part of the total spectrum of electromagnetic radiation is not necessary for photosynthesis. This is the better quality PE-film protected against the influence of UV radiation. The durability of such film is three to four years. This film is recommendable for all kind of greenhouses irrespective of climatic condition. The thickness of these types of films ranges from 100 micron or less to 200 micron (μ) or more. 100μ films are used for small or tunnel structure for shorter duration. 150μ or 200μ films are generally used for covering, 250 or 300p are used for partition and apron or skirting respectively.

3. PE-IR film: IR-films were introduced in 1983 for conserving more heat, thus saving energy for heating. It can reduce transmissivity of long wave radiation (IR). This film is recommended in subtropical highlands and other places of low night temperature, where the ventilators are closed down at night. This particular property of this film can trap more radiant heat inside the greenhouse thus increasing the inside temperature than the outside at night. It is generally 0.2 mm thick and may be UV stabilized. The light diffusion character of IR-film is even more beneficial to plant production. In addition, when going from a cloudy day to a clear/sunny day, the IR/AC (thermal) film helps to speed up transpiration.

4. Mono-layer UV and anti-drip PE-film: This is 200 micron thick transparent PE film which is UV-stabilized. Some such films are at the same time have characteristics that does not allow the condensed water layer or droplet formed in the lower surface of the film to fall on the crop as drops, after coalescing. Instead this character of the film allows the condensed water to run down, as water film, along the slope of the roof, preventing to form a drop. The micro drops in the PE sheets reduce the light transmission in the early hours of the day and are harmful to the crops when they fall on them from roof and also induce disease infestation. In the areas with high humid winter season these PE-anti drop films are most suitable.

Mono-layer Bubble film: These greenhouse covers are generally made up of monolayer films, having 200µ thickness, made of LDPE and poly vinyl acetate (PVA) with air bubbles entrapped in the middle. Due to the air bubble in the middle, this film is more efficient in maintaining the temperature in the greenhouse.

5. Multi layer/character PE film: In recent times new types of multi layer PE-films have come into market, which have different characteristics in different layers in their 200µ thickness. Apart from characteristics like UV-stabilized, anti-drip, there are few other characteristics like anti-dust, anti-mist, anti-sulfur etc. It is technically superior and cheaper in respect of the advantages it provides to the greenhouse. For example a 200 micron film may be produced by combining two layers measuring 120 and 80 microns each, of which the 1st or upper one is UV resistant and the second one have anti-drop character. Reduction of linear-expansion coefficient is also available with this type of film. Thus, three-layer and five-layer films are developed and used to protect crops from hail, wind, snow, or strong rainfall. In horticulture, together with the realization of a confined space, with controlled microclimatic conditions to maintain favourable conditions of temperature, solar radiation, gas composition (O_2, CO_2, and N_2), and humidity we can induce significantly the optimal photosynthesis, respiration, and subsequent growth of the plants.

6. Diffuse or white PE-film: It is a white colour film, instead of clear film. The incident light can reach the plant as "direct" radiation (waves moving parallel to one another) or as "diffused" radiation (in which the light waves are moving along trajectories which are at various angles with respect to one another and to a horizontal plane). In general the PE-films are clear in nature and allow the "direct" sun-light (80-90%) in to the greenhouse. Diffused PE-film allows the sun-light in a diffused form. Due to the light diffusion there is no shadow inside when a greenhouse is covered with diffused film. If there are hanging baskets and plants on a bench underneath, they both get the same amount of light.

7. Woven PE sheet/cloth: It is tear-resistant, woven cloth made up of polyethylene tape. The Woven Greenhouse Film is designed for long-lasting and beneficial coverage. The thickness of such film is expressed as mil (0.001 inch) and in general 10mil is considered for greenhouse. It has many advantages in comparison to PE-film. It is strong (high breaking & tear strength that protect it from hail), diffused, and lasts long. It is UV-stabilized and can have the other characters like anti-drip etc. The speciality of this film is its high tear and impact resistant character, which makes it suitable for stormy tropical areas. The two negative aspects of this film are its transmissivity, which is little less (around 80%) but suitable in tropical climate, and its cost, which is little more than the PE-film.

4.3.2. Non-transparent and water-permeable materials

Two completely different groups of such materials are used generally to construct shade-house or lath-house (a) Non-plastic or dried plant materials, (b) Plastic nets.

(a) Non-plastic or dried plant materials: It is being used for few specific crops since long. For smaller areas, the dried twig, leaves (coconut, banana etc.), sticks, grass, straw, jute stick etc. are used to cover the structure constructed with wood, bamboo, wire/rope etc. under which the cultivation is done. Generally the shade loving tropical crops are grown in such greenhouses. It saves the crop from impact of heavy rain and wind and also maintains a steady humidity status apart from giving necessary shade. However, the problems with these structures are:

- The durability of such covering is about one year.
- Maintenance of such structure is difficult.
- Large area may not be feasible to maintain with such covering.
- Nowadays availability of such plant material is uncertain.
- No control over the percentage of shade provided to the crop.
- May act as the host of disease and pests and may attract harmful microorganisms at the time of degradation of these organic materials.

Nowadays no promoter plans for such greenhouse except the people of a few tropical countries like India who are still growing betel-leaf under such shade-house.

In cool climate or low temperature (below 0°C) winter, especially in snow falling areas, sometimes the shade house is covered with thin wood strips (about 5 cm 2 inches wide) to give one-third to two- third cover depending upon the need of percentage of shade.

(b) Plastic net: It is mostly the PE-shade net. However, insect-proof nets are also used for greenhouse construction, either as supporting covering material or principle covering materials (Insect net house).

There are a number of plastic materials prepared like a net to cover the shade house. Some commercially available materials include UV-stabilized cross woven polyethylene fabric that resists ripping and tearing. Apart from this, UV-stabilized high-density polyethylene (HDPE) shade cloth knitted according to the requirement, is strong and had a greater longevity. The shade cloth is resistant to ripping, and has an optimum life of 10 to 15 years, depending upon the climate and quality of the material. It is available in rolls of 50 m having a width of 3 m or any other size. These are of different colours, generally black, green, white, and red which are used for different climatic conditions, different solar radiation (irradiance) and for different crops. These shade nets are woven in such a way that specific knitting passes only a fixed percentage of light through it. On the other hand it can provide a fixed percentage of shade to the crop. This shading values range from 28% to 75% (Table-7). In most of the cases 50% is recommended for covering a general shade house to manipulate the solar radiation without hampering the growth. However, for shade loving plants or for nurseries up to 75% shade nets are used. To cut solar radiation in PE-film greenhouse, to stop the mid-day temporary wilting of crops or to reduce evapotranspiration, the lower percentage shade nets (25% to 50%) are used below the roof of greenhouse. Examples of utilization of HDPE UV-stabilized shade nets are given below.

The insect-nets are of similar type but not used to provide shade but to protect crops from insect pest. The porosity of these nets is less than shade-net and expressed as mesh. It is commonly of 40/50/60/70 mesh depending on the targeted insect/s and is UV-stabilized.

Table 8: Radiation status and use of different types of HDPE uv-stabiized shade-net

S. No.	Radiation	Crops/Purpose	Shade net	
			%	Colour
1	High solar radiation	Indoor plants, cardamom, etc.	75	Black
2	Moderate-solar radiation	Orchids, Anthurium, nursery materials etc.	50	Black/green
3	High solar radiation	Gerbera (use under PE-film) and some vegetables etc.	50	Black
4	Moderate solar radiation	Flowering plants	50	White
5	Any situation	Any shading purpose including wind break	50	Green

The shade house can also be built with movable shading PE-net. In case of low incidence of sunlight when crops do not need shade the shading material can be pulled aside.

Shade net as windbreaker: Sometimes artificial windbreaker is created with the green shade-nets. In that case the porosity of the nets will represent the porosity of the windbreak.

4.3.3. Fixing of covering materials through fastening and stretching

To fix the covering materials on the structures of greenhouse, PE-films or PH-nets must be fastened and stretched on the structural elements of construction. The PE-film, in particular, have to be fastened with technically equipped methodology, otherwise the films may tear apart from the point of fastening while pressure is put on the film. At the same time the PE-films have to be stretched tightly to prevent them flattening in the wind so that they are not destroyed.

The shade-net has the advantage of avoiding wind and water pressure to a greater extent. So the fastening and stretching of PE-shade nets are easier in comparison. The method of fixing of covering materials should require small number of workers and should take short time. Now the methods of fastening and stretching of both polyethylene film and net are detailed below separately.

4.3.3.1. Fastening and stretching of PE-films

The PE-film is first fastened at one end or both the ends, generally along the length, and then it is stretched. From simple nailing to technical fastening methods are followed in different types of greenhouses.

i.   One simple method is that the film is nailed on the wooden frame with wooden slats. In this case the stretching is done manually with a limitation. It is a time consuming operation.

ii. It is better to screw the film between wooden slats or between steel construction and wooden slats: It is easier to remove. Some constructions have wires with which the film is clumped together. The use of nail and screw are generally feasible for relatively small greenhouses with larger area of permanent non-closable ventilators along with rollup curtains. Such structures require less stretching in comparison to tall big greenhouses with ventilators only on the top or ridge.

In the structural frame, made up of steel or galvanized pipe, the PE-film can be fastened at the upper edge of the roof with plastic clips. A very good method of stretching of films on such construction is to fix the film on a steel pipe/tube on the gutter and then fixing that tube in the gutter profile (Fig. 24 a). In this method the film is fixed to the pipe placed on the gutter with adhesive tape or small clips. Then the pipe is turned with a handle until it is positioned in the curved part of the holder. The pipes will be fixed at the end and the film will be stretched tightly. These methods of fastening and stretching of the PE-film can be done with pipes up to 40m long.

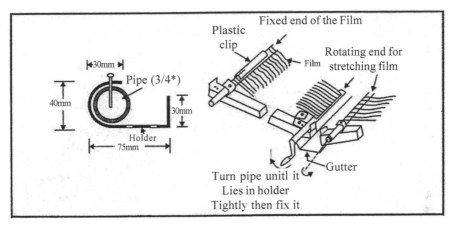

Fig. 24 (a): Stretching and fixing of PE-film

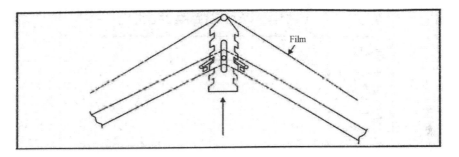

Fig. 24(b): Mechanical stretching device of the film

(iv) Instead of plastic clip made up from plastic pipe many others fastening devices (Fig. 25) made from plastic, aluminum and steels are available with the modern greenhouse contractors. These fastening devices are set firmly in the required position of the structure. The film can be fixed firmly or taken off easily from the construction without damaging the film. The film can be stretched mechanically or manually after fastening one end. The fastening device at the other end can be fixed when the necessary stretching of the film is completed.

There are a few other methods of fixing/stretching the PE-film after fastening.

- Use of hand or manual stretching (it is most common).
- Using plastic rope stretched over the construction.

- Using mechanical stretch devices.
- Using wooden sticks that are placed alternatively on and below the film.
- Using rolling up of the film on a steel pipe in the gutter.
- Using load (water filled pipes) along the length, and is also used as gutter.

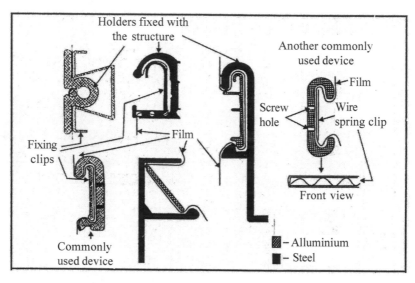

Fig. 25 : Different types of Fastening devices for plastic film.

4.3.3.2. Fastening and Stretching of PE-shade net

Fastening and stretching of PE-shade net is far easier than that of PE-film. It can be simply attached to heavy wires/ropes fastened to supporting posts and stretched manually before fixing the ends of the net just like tarpaulin. However, the manufacturer of HDPE shade-nets use to perforate the nets at the edges and sometimes provide pellets that make the stitching of two nets more easy (Fig. 26). For proper stretching, clips may be used with rope (Fig. 26). At the finished end of the net or at the heat-cut end of the net the clips may be fixed from both side of the net-end along with a rope. If scissors is not use to cut the net then fold the cut selvedge and then fix the clips. The clips may be used at every 50 cm along the side of the net to stretch or fasten the net against the structure. For proper stitching of two nets, pellet may be used which are stitched with a plastic rope made up of UV-stabilized HDPE materials (Fig. 26). The pellet is placed through each aligning eyelets of two nets as shown in the figure. Then the rope is passed through the pellets and fastened with the structure.

Pellet

Fig. 26: System of stretching and fixing of shade net with clips & pellets.

4.3.4. Insect Proof (IP) nets

Insect proof net is almost essential nowadays particularly in the greenhouses having ventilation system for cooling purposes. Plastic or polyethylene nets of different mesh size are used to cover the openings of the greenhouses to protect the crop from smaller pests. In general IP-nets of 50 mesh is used to cover the openings.

The application of PE-IP-nets reduces the ventilation efficiency and increases the temperature and humidity of the greenhouse. IP nets with different mesh sizes are available. Decrease in mesh size increases the inside temperature and humidity.

Experiment of Prof. V.M. Salokhe shows that the greenhouses having large openings with 78 mesh had around 1-2°C (average) higher temperature than the 40 mesh. Similarly the 40 mesh raises temperature up to the same magnitude against 34 mesh. In general 50 mesh IP net is widely used to protect the openings from entry of insect, and as increase of inside temperature by increasing the mesh size is small, the 78 mesh may be used to screen the very small insects like thrips.

However, according to Harmanto and V M Salokhe (2006) nets of different mesh size had a significant effect on microclimate and air exchange rate. Compared to 40 mesh greenhouse, the reduction of air exchange rate of about 50% and 35% was obtained for the 78 and 52 mesh greenhouse respectively. At the same time the position of absolute humidity in the greenhouse had a pronounced effect with the change in mesh size of the IP nets. Humidity in 78-mesh greenhouse was consistently approximately two times higher than that of 40 mesh greenhouse, while 50% increase is observed with 52-mesh greenhouse.

4.4. Construction Particulars

4.4.1. Orientation of greenhouse

It is an important aspect of greenhouse construction. The proper position of roof covered with transparent material to invite maximum transmission of sun-ray inside and proper position of structural frames to avoid casting of shadows are the two important factors to be considered for orientation of a greenhouse. However, the changing position and angle of sun is a problem while finalizing the proper orientation. The problem is more serious in low-radian i.e. in cooler areas normally situated beyond 40° latitude.

In tropical areas the problem of availability of solar radiation is less, but the problems of cloudy sky during rainy season are there. In tropical region effective ventilation is important to control the high temperature and humidity by way of proper orientation of greenhouse.

The single stand-alone greenhouse located in cooler regions may be oriented east-to- west, which will provide less shading. This will also allow the low angle sunlight during the beginning and end of the day (Table 9).

In warmer regions the single stand-alone greenhouse shall be oriented from north to south as the angle of sun-ray is much higher. This also resists the wind pressure flow from north to south efficiently. However, in case of naturally ventilated greenhouse, the orientation may be east to west, which accentuates the effectiveness of roof ventilation.

Table 9: Effect of orientation of a single greenhouse on sunlight transmission at latitude of 50°N (calculated by L G Morris).

Orientation	% of transmission of sunlight inside	
	Mid-summer	Mid-winter
North-to-South	64	48
East-to-West	66	71

In gutter-connected hot climate greenhouses the house shall be oriented north-to-south where the gables are automatically placed east-to-west. Here the natural ventilation system will work better using the flow of wind directed from south in summer and from north in winter (Fig. 19 & 20).

On the other hand in cool climatic regions the gables will be placed north-to-south that permits the shadow of north roof and glitter to move across the floor during the daytime thus shadowing in a particular place can be avoided. The transmissivity is not dependent on number of span, but it decreases if the ratio of the length to width is greater than 5.

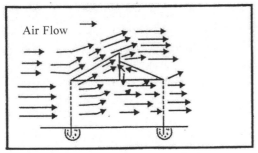
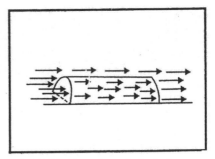
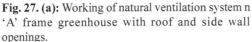

Fig. 27. (a): Working of natural ventilation system n 'A' frame greenhouse with roof and side wall openings.

Fig. 27 (b): Natural ventilation in Quonset type greenhouse

4.4.2. Design of load

Calculation of the total load of a greenhouse is essential for designing and construction of a greenhouse. The structure should be capable to withstand weight, up to certain limit either vertically or horizontally before collapsing. On the basis of total load of a greenhouse the type of foundation, the quality and section of posts and structure of roof will be determined. The total load of a greenhouse generally includes dead load, live load, and wind as well as rain/snow load. These loads are estimated on average basis but the extreme situation should be considered before finalizing the structural design. On an average a poly-greenhouse of non snow fall zones will be designed to bear a minimum load of 25 kg/m² and to resist wind up to 150 km/hr. However, depending upon the actual live load and wind speed the calculation shall be done.

Dead load: It is the weight of all structural material used for construction as well as the weight of all kind of instruments attached to the structure. It includes weight of roof frame including gutter, covering material, pillars or posts, fans and lights etc. This load for a particular greenhouse is almost constant. In case of steel frame structure the average requirement of MS steel is about 4.5 to 5.5 kg/m².

Live load: This generally includes (1) load of hanged or trellising plants, (2) load of men for any work on the structure, and (3) accessories required for these purposes. For cultivation of trellising vegetables this load factor is vital for structural design of a greenhouse. Frequently tons of crop-load is required to be borne by the structure. Thus additional crop load shall be predicted properly. An average estimate of crop load for tomatoes and cucumber in a high-tech greenhouse may be considered 20 kg/m².

Rain and snow load: They are the two different types of loads that can be estimated from the analysis of climatic data of the area. In case of cooler region snow load is an important factor for structural design as in some places it is as high as 75 kg/m². In case of warmer and humid region the load of rain splash is required to-be considered but it is nominal in comparison to snow-load. Both these loads are applied vertically from the horizontal projection of roof. The adjustment of roof angle can reduce the rain or snow load of a structure. For rain, higher slope is not essential but for snow the slope should be more than 25 to 30° to encourage sliding. For no-snowfall areas the slope may be reduced to 15°.

Wind load: It is the load created by wind blowing horizontally and transferred to the greenhouse structure. In naturally ventilated greenhouse this wind load sometimes create an upward thrust. To estimate the average peak wind speed of an area, data of 25 years mean recurrent interval may be used to calculate the wind load. The effective pressure of wind can be calculated from the wind speed using the following formula.

$$Q = 2.37 \times V / 10^5$$

Where q = pressure in Pascal, and V -basic wind speed in m/s.

BOX - 11

Wind load calculation

To calculate wind load some basic components of wind load is to be considered: The effective velocity pressure can be calculated by converting wind speed using the following formula (ANSI A 58.21- 1972 6.3.4)

$$q = 2.37 \times 10^{-5} V,$$

Where q - basic wind pressure in Pa (Pascal) & V = basic wind velocity in m/s.

Other factors such as gusts and height facility can be included in the final calculation; The final wind load factor (w), for greenhouse is calculated by the following formula.

$$W = q \times q_P,$$

Where q_p = external pressure acting at local position.

The external pressure coefficient have been taken out from tables (ANSI,1972) and shown below as positive and negative pressure acting on three common type of greenhouse structure: considering the wind speed 30 m/sec.

WIND DIRECTION

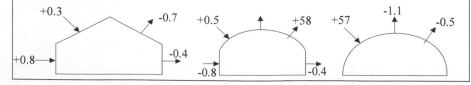

4.4.3. Constructional details

Construction and fabrication of structural members of a greenhouse is important mainly in relation to their load bearing capacity apart from other climate controlling aspects. Structure of a greenhouse consists of four distinct units viz. (1) load bearing foundation, (2) load transferring pillars/posts, (3) roof structure and (4) structures for door and climate control machines. After introduction of polyethylene film instead of glass the structural design and construction of greenhouse has been simplified. Reduction of roof load is the main cause of this simplification.

4.4.3.1. Foundation

Foundation is the base of any structure. This bears the total load of the structure transmitting it to the soil. Foundation is an important aspect of a greenhouse that provides proper anchorage to withstand both vertical and horizontal load. Out of all the different components of loads only wind can make a big difference because of its high variability in comparison to other loads. Apart from average normal speed of wind considered for a greenhouse, storms or strong gusty wind for few hours is very common in many areas and can destroy a greenhouse because of its low ratio of weight to surface area. Foundation can play a great role to avoid such incidences.

Several types of foundations are used for greenhouse construction; however, out of these, only two to three types are considered for commercial greenhouse construction.

(a) Natural /minimal foundation: It is the simplest and shallow foundation where the pillars or posts are placed directly on or in the undisturbed supporting soil. This type of foundation is generally used in relatively smaller greenhouses with lower height having low wind load. The posts of such greenhouse directly pass on the total dead load and live load to the ground firmly compacted to withstand the load. In this foundation the posts are sometimes placed on stone slab of concrete footings placed in the soil at least 24 inch below the ground surface.

The wind load, which acts horizontally, creates enormous side and upward pressure that generates an upward thrust to the foundation. In high wind-speed areas the foundation should be firmly fixed in the ground to avoid toppling or displacing of the structure. In such areas frequently the supporting guy-rods or frames or wires are used to protect the structure from strong wind and naturally the foundation of these supporting structures needs proper supervision during construction.

(b) Wooden foundation: In earlier days, this easy and simple way to construct the foundation of small and hobby greenhouse was popular. Presently it is not used extensively due to scarcity of quality wood. However, resistant wood of 4 inch X 6 inch size (4X4 for small size) timbers may be used for foundation. The timbers are laid with proper leveling on the ground, on which the greenhouse will be set. If the root development permits, polyethylene sheet may be used to cover the entire floor area below the timber frame to control weed and capillary rise of excess water. The corners of the rectangular wooden foundation base of the greenhouse are properly fixed. If necessary this foundation frame may be fixed into the ground with nail or spike like structure. Then the frame of greenhouse structure is fixed on it.

(c) Concrete foundation: It is the simplest and cheapest form of concrete foundation used at the base of the posts/pillars. It is also known as direct block foundation for greenhouse.

However, several more useful and indirect concrete foundations starting from simplest and cheapest concrete block-foundation to complex and costly slab-foundation are being used for greenhouse depending upon the purpose and size of the project. Such types of foundations are narrated below.

(i) Concrete Block foundation- It is the most common and practiced type of foundation used by poly-greenhouse or shade-house. It is basically of two types, **(a) Concrete block reinforced with foundation pipe** (telescopic arrangement) on which the pipe of the pillar/post is fixed / bolted. It is the wide-spread method adopted by almost all the poly-greenhouse where the structure is not heavy like glass-greenhouses. **(b) Concrete Block with anchors & foundation bolt** on which the posts/pillars is fixed. It is only used for heavy commercial greenhouse structures or for any specialized design.

Pre-cast foundation- It is a new concept, innovated by the author, to execute concrete block foundation. Basically it is an operational modification of a type i.e. concrete block reinforced with foundation pipe mentioned above. It is precast at a suitable place then carried in the project site and buried in foundation pit to make a foundation. It saves cost, time and give better strength. (Photo 28).

(ii) Concrete wall foundation- This type of foundation gives good support for heavier structure like glass-greenhouse. The footing of the wall is constructed at the recommended depth of the trench made into the soil. The footing is usually twice as wide as the wall. After hardening, the wall is constructed up to the height of the drain. The anchor bolts are fixed at the top of the concrete wall with a spacing of 4 feet starting one foot away from each corner. A 2"X 4" sill made up of wood or plastic composite lumber may be used as *base* on the top of the foundation and is attached to the greenhouse using the concrete anchor bolts (Fig. 29 C).

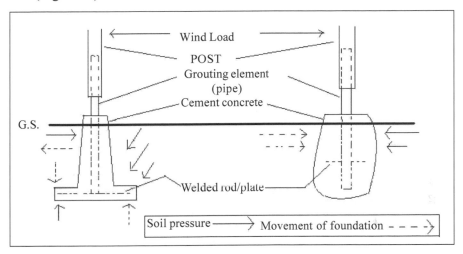

Fig. 28: Design and efficacy of pre-cast grouting system

(iii) Concrete slab foundation: A concrete slab makes a convenient base for a commercial gutter connected greenhouse where the grade of finished floor should be levelled one step below the floor of the house. For stand-alone greenhouse the floor should be several inches above the finished outside grade of the floor. The Concrete slab should be 1 inch wider and longer than the outside dimensions of the greenhouse. A 3 inch thick floor is adequate for small greenhouse (Fig. 29 D). The outer edges should be thicker to give support and resist cracking. The inside drainage system should be provided within the slab. Mulch shall never be used, as this will become a nest for bugs and pests that will run havoc in the greenhouse. In very moist soil at least 4 inch of compacted gravel or stone may be placed on the top of the sub-soil and covered with 6mm polyethylene film that will keep the slab dry.

(d) Foundation for Bamboo greenhouse: Recently introduced innovative and scientific bamboo greenhouse (innovated by the Author) requires proper foundation. Here, one extra measure is necessary to be taken when grouting system of bamboo posts are to be designed. The bamboo should not be in direct touch with the soil. Two types of grouting system is recommended, 1st is direct grouting system of concrete foundation, and 2nd is indirect grouting system of steel-concrete foundation. In both the cases one have to take care of suitable and proper measures to make these two types of foundations strong and to keep the bamboo-posts separated from soil. (For detail see chapter 5; the Bamboo greenhouse)

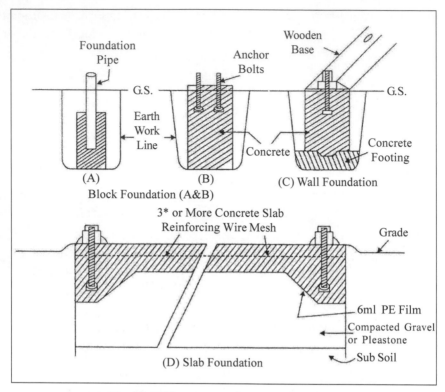

Fig. 29: Different types of concrete foundation

The following steps are to be taken to construct these types of foundations.

1. After leveling of the land the position of pillars should be marked according to the design of the greenhouse.

2. Earth cutting of the pits shall be done. Depending upon the span of greenhouse and height of the structure the pit size is determined. The diameter of the pit may be 0.23 to 0.45 m with the span of 8 to 20 m respectively. The diameter also depends on spacing of the pits. However, the depth of the foundation pit should be minimum 0.60 m (Table 10). However, in the areas where frosting occurs, the foundation should be below the frost level.

3. Depending upon the size of greenhouse and type of pillar any of the three types of block foundation system can be adopted. First the foundation pipes, just bigger than the size of pillars can be fixed in a concrete block of required size constructed inside the pit and below the soil surface. The pillars will be inserted into the foundation pipe. In case of Quonset or 'D' shaped structure this is a common practice.

Table 10: Relationship between pit diameter (m) and span of a greenhouse to ensure proper strength of the structure.

Greenhouse span (m)	Pit spacing (m)					
	1	2	2.6	3	3.3	4.5
6.6	0.15	0.23	0.30	0.30	0.30	0.38
8.0	0.23	0.23	0.30	0.30	0.30	0.38
9.3	0.23	0.30	0.30	0.38	0.38	0.45
10.6	0.23	0.30	0.30	0.38	0.38	0.45
12	0.23	0.30	0.38	0.38	0.45	-
13.3	0.30	0.30	0.38	0.38	0.45	-
15.3	0.30	0.38	0.38	0.45	0.45	-
20	0.30	0.45	0.45	0.45	-	-

If the greenhouse is somewhat bigger in size, the pillars are not put into the foundation pipes. Instead, the concrete block foundation will be constructed up to a little height above the ground surface with bolts fixed on the top of it. The base-structure of the pillars has necessary holes that are bolted with the nuts of the concrete block foundation (Fig. 29 B & C).

4.4.3.2. Pillars or posts and roof structure

These are the vertical components for carrying the greenhouse structure. The pillars or posts of a greenhouse are the main supporting structures that bear the entire load of the greenhouse. Properly designed pillars can efficiently bear the vertical load of the structure except the wind load, which is uncertain and sometimes creates serious problem. Angular support to the pillars is necessary to resist this. The load bearing capacity of the pillars depends on the type of material, size and cross-section of the members (Table 11).

Table. 11: Average weight and live load factor of materials of different size and shape required for construction of greenhouse

Cross section	Material	Size (mm)	Weight (kg/m)	Load factor
L - (angle)	Steel	65 x 65 x 5	59	40
[- (Channel)	Steel	160 x 100 x 4	295	200
| - (I-beam)	Steel	50 x 80 x 4	171	116
O - (Pipe)	a) STD Steel	32(OD)	37	-
	b) H T Steel	32(OD)	89	60
	Tubular steel	50 (mm)^2	111	75
	Formed steel	—	44	30
Rectangle timber	Wood	(a) 50 x 100 (fir)	37	25
		(b) 50 x 150 (fir)	77	52

Different types of steel, aluminum, reinforced concrete pillars, woods, bamboo etc are generally used for construction of structural frame.

Reinforced concrete pillars and bamboo are generally used for 'low-tech' and shelter type greenhouse generally found in countries located near equatorial areas. Wood and bamboo cannot withstand high humidity and decay early. Non cured and untreated bamboo is not commonly recommended for constructionof normal greenhouse. Thus matured and seaseoned/cured bambo is recommended after it is treated and painted for protection from high moisture and termites. The chemicals used for treatment and paint should not be hazardous to crop or human health.

Due to poor load bearing capacity and high cost aluminum is not generally recommended for construction of greenhouse. However, for high-tech greenhouses many of the accessory items and members are made up of aluminum. Light weight and less affected by moisture are the two reasons for which it is sometimes preferred over steel.

In general main structural members, particularly the pillars, used for greenhouse are made up of steel due to its load bearing capability and durability. However, rusting in moist condition is the .main problem of steel structure. The best way to prevent this problem is the hot-deep galvanization of the steel-frame members.

Specification and requirement of structural members/components made up of galvanized steel

Columns, Purlins, Truss bottoms, Roof top-chords, End frames, Gutter purlins, Hockey / column support, Bracing / strut, and other steel made structural materials required for curtain, flap, and different climate control apparatus, are the structural members, commonly required to construct a greenhouse.

Size and specification of these structural members normally depend on (1) Wind pattern of the area, (2) Load calculation, (3) Spacing of columns, (4) size of bay, and finally (5) design of greenhouse including the status of ventilation area.

Unfortunately very little information is available in these aspects, particularly in respect of greenhouse suitable for subtropical humid climatic situations. However, depending upon the common practices and through some trial and error-work done in different areas, an average estimate of size and specification of different structural members is made, which is narrated below.

In the table below only the galvanized A-class pipes made up of 2mm thickness steel has been considered. The outside diameter (Ø) is mentioned in millimeter (mm).

Table.12: Greenhouse-wise requirement of structural members.

Structural member	Stand alone A-frame type structure		Quonset type structure		Gutter-connected saw tooth type structure	
	High wind	Low wind	4-6m wide	8-9m wide	High wind	Low wind
1. Column / hoop	50	40	25	40 ?	76 & 50	50
2. Purlin	25	25	15	32 ?	32	32
3. Roof top chord	25	25	-	-	42	32
4. Roof bottom	32	25	-	-	50	42
5. Bracing/strut	-	-	-	-	32	25
6. End frame	25	25	-	-	32	25
7. Hockey	32 & 25	-	15	25 0	50 & 42	32
8. Other structure	25	25	15	15	32	32 & 25

It is also estimated that in a saw-tooth type structure, on an average, the requirement of steel is roughly about 5.5 kg/m² i.e. 5.5 metric tonne per 1000 m² size greenhouse.

Table 13: Approximate materials (pipe & wood) required for a low-cost 100m² mixed structure shown in Fig.15.A

Item no.	Material	Section (mm)	Length (m)	Number
1.	Steel pipe	40?	2.5	18
2.	Steel pipe	40?	3.5	9
3.	Steel pipe	32?	5.0	9
4, 5, & 6	Wood	50x75	2.5	32
7.	Wood	50x50	2.75	9
8.	Wood	50x50	3.0	9
9 & 12	Wood	38x50	2.5	32
10.	Wood	38x38	3.0	8
11.	Wood	38x38	2,75	8
13.	Wood	38x50	5.0	2
14.	Steel pipe For foundation	50 ?	0.75	27

4.4.3.3. Door

It is an important part of a greenhouse which is the only entry point. Entry of men through this structure is a vital aspect to manage the operation of the greenhouse cultivation system.

Risk factor- Interestingly this is also the main entry point of pest and other organisms which infest the inside crop. Thus door should be designed properly and also operate accordingly. For smaller greenhouse one door is sufficient but for larger and multi chambered greenhouse two doors (of which one may be bigger) are often constructed in different places for entry passage. However, most of the operations are normally done through one door.

Two types of doors are generally used for greenhouse, i.e. 1) single door flashing to the sidewall of the greenhouse and 2) box type double-door fitted with two sliding doors and the box fitted in sidewall (Fig. 30).

In case of low cost stand alone greenhouse single door is commonly used. Split curtain may be provided with these doors to avoid free entry of pests at the time of opening of the door for operation.

For better quality greenhouse meant for quality production the box-type door is commonly used. Generally sliding panels are used as door, and two such doors are fitted in opposite walls of the box (Fig.30). This will ensure protection from entry of the pests. It may be used for both stand-alone and gutter-connected greenhouses.

Size of the door as well as box is decided depending on the size of the greenhouse providing a fair space for a man with material to enter.

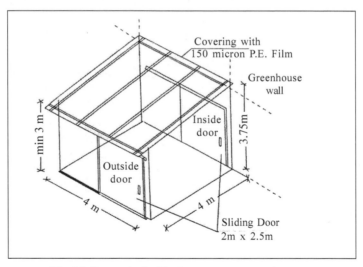

Fig. 30: Box type double door structure greenhouse

Rigid plastic sheet, polyethylene film and insect-proof net are generally used to cover the door. However, the roof of the box-type door should be covered with non-permeable materials like rigid plastic sheet or PE-film to protect the floor of the box from rain.

The frame of the door is commonly constructed with quality wood and/or galvanized steel. In case of wood it should be painted properly with quality paint.

The box door is also used to house the electric control panel and different tool boxes.

4.5. Operation for Maintenance of Structure of Greenhouse

Greenhouse cultivation is probably the most capital-intensive method of cultivation. It is, therefore, necessary that a greenhouse shall be utilized in a proper manner and maintained regularly to increase the efficacy and lengthen the life of structural membranes, covering materials and different parts.

First, it is about the maintenance of greenhouse structure. Here the life of plastic covering material (3-5 yrs) is always less than the life of the structure, and this provides a good scope of thorough maintenance of the structure after every 3-5 years.

For better performance and longer life the maintenance of covering material and different parts of greenhouse can be divided in the following segments.

1. Structural frame and door: In case of wooden structure the posts and other wood in contact with ground should be treated properly before use to prevent any damage due to weathering. It is better to go for readymade treated wood for these purposes. In such structure all wooden parts of the structure require continual painting for protection against rot, particularly in moist and high humid environment. However, mercury based paints must be avoided. In respect of colour, white paint is most suitable. In spite of protective coat of paint, eventual replacement of ratted portion of the wood / frame is necessary.

    In case of steel structure the pre-fabricated structural material or basic structure materials are to be galvanized properly (0.2 mm thick). In case of non galvanized mild steel (M S) structure, the structural materials should be painted with A-class paints (silver or white colour) every two year.

2. The wheels of doors and the ridge, on which the wheels move, should be greased properly on monthly or bi-monthly basis.

    In case of automatic ventilator/window operating systems, the racks, adopters, motors, and gears should be greased monthly. The manual system also needs proper greasing regularly as per requirement.

3. Cleaning of covering materials: The accumulation of dust particles on the roof should be removed on monthly or bi-monthly basis. However it should be done with special care before onset of the rainy/cloudy season. The cleaning can be done with plain water. The monthly cleaning removes the algae frequently grown and spread on the covering material particularly in prolonged rainy season.

4. When white (lime) wash is done on the roof covering plastic to protect crops from strong radiation in summer, normally most of it wears off by the first rain; if not, it needs to be washed off.

Maintenance of Structure of Bamboo Greenhouse: It is really very important to have a proper schedule of maintenance of Wood and bamboo greenhouses. Four types of maintenance schedule is commonly recommended for such greenhouses. These a1re 1) 1st year maintenance schedule, 2) Yearly maintenance, 3) emergency maintenance, and 4) thorough maintenance with film replacement. The details of this maintenance are stated afterwards in the bamboo greenhouse chapter (chapter 5).

4.6. Details of Plastic Tunnels

4.6.1. Low plastic tunnels

It may be defined as "the low height structure covered by any transparent material or net made up of plastic to envelope only a row or bed to enhance the plant growth under it". (See Fig. 21.a)

Primarily transparent plastic films are used for this structure, which increases the inside temperature around the plant, thus helping the summer crop to grow better in winter. It also advances the maturity of crop by 30 to 45 days. However, where the temperature difference between summer and winter is not so pronounced and where the average night temperature in winter season does not drop below 10°C, this structure does not yield the expected result.

It is a simple structure where covering material may be supported by hoops, made up of rod or pipe, closely placed over the crop for a relatively short period of time (2 to 8 weeks) depending upon the crop and the climate. This low plastic tunnel is sometimes used immediately after transplanting of seedling to boost its early vegetative growth in the non-supportive cool ambient temperature. Cover shall be removed when plants begin to flower to allow necessary pollination. The added cost due to this structure is recovered profitably through higher off-season price.

Crops: In general the low plastic tunnel, covered by PE-film, are used to cultivate cucurbitaceous vegetables like bottle gourd, bitter gourd, cucumber, summer squash, water melon, mask melon etc. Almost all such crops can be transplanted in winter under this structure and thus the harvest can be advanced by 25 to 50 days in North Indian condition (Table 14).

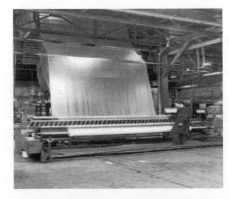

PE- film production

Manual pipe-bender for bending of pipe

Greenhouse structure manufacturing work-shade

Stretching _ fixing of PE-film

Structure for nursery purposes

Large 1 acre greenhouse

Handle to operate shade net

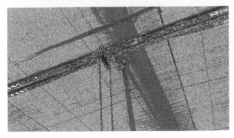

Chain sproket to remove shade net

Different roof structure

Different roof structure

Different roof structure

Urban greenhouse

Low-cost small greenhouse

Innoveted precast grout for greenhouse

Table 14: Time of transplanting and advancement of harvest of different crops grown under low plastic tunnels in North Indian condition.

S. No	Crop	Time of transplanting	Advancement of harvest
1.	Bottle gourd	1st week of January to 1st week of February	30 to 40 days
2.	Bitter gourd	3rd week of January to 1st week of February	25 to 45 days
3.	Cucumber	4th week of January to 1st week of February	30 to 35 days
4.	Summer squash	1st week of December to 4th week of January	30 to 60 days
5.	Long melon	1st week of January to 1st week of February	40 to 50 days
6.	Water melon	2nd week of January to 1st week of February	30 to 40 days
7.	Musk melon	2nd week of January to 1st week of February	30 to 40 days

Source: Indo-lsrael project, IARI, Pusa, New Delhi.

Low plastic tunnels with other type of covering materials: Other covering materials like PE-shade-net, PE-IP-net, may be used to cover low plastic tunnels for growing crops primarily for different reasons.

Low plastic tunnels with shade-net are used to reduce high solar radiation during daytime, increase humidity and provide high temperature at night to the crop grown under it. This also protects crops from hail, heavy rain, birds etc. This structure is suitable in the areas where average night temperature in winter does not go below 10°C and where the difference between day and night temperatures in winter is wide.

The low plastic tunnels covered with PE-IP-net is almost similar except that it restricts the entry of the destructive small pests like aphids, thrips, white flies etc. Since most of these pests act as vectors, restriction to their entry also restricts the infestation of many viral diseases. In tropical area it will be an important aspect of cultivation, in due course.

These two types of structures are best suitable for raising seedling or other plant materials in off-season and/or in disease free condition.

Specification of low plastic tunnels: This structure have arch like supporting frames, termed as hoops, which are generally made of flexible galvanized iron (G) rods (6 to 8 mm) or plastic pipes (12mm) (Fig. 21a). The ends of these hoops are fixed in the soil at the both sides of the row or bed with the spacing of 1.5 to 2 m. The height and the width of the hoop at the ground level will be equal. This is generally maintained at 45 to 100 cm.

The rows or beds covered by the tunnel shall be laid at North to South direction to receive maximum sunlight. Transparent PE-film of 30 to 50 micron or shade-net / IP-net of desired perforation are commonly used to create the tunnel. GI or plastic threads or ropes are sometimes used to fix the film/net from outside along the position of hoops to protect the tunnel from strong wind. The ends of the plastic covering are normally buried in the soil thus providing necessary stretching. Finally, from outside, the tunnel looks as an elongated white coloured bund-like structure lying horizontally over the field. Series of such tunnel-structures with a gap in between shall cover the entire field. The hoops supporting the structures from inside externally looks like the ribs of these structures.

However, depending upon the climate and crop, the height of the tunnel may be reduced further. In that case the height of the tunnel may be half of the width.

The crop rows are placed in the middle of the bed and the two end points of the hoops are placed equidistant from the row. Drip lines i.e. laterals are placed along the rows before transplanting and each seedling are placed very close to each emitter (Fig. 52).

Operation: The plastic is usually covered in the afternoon after transplanting has been done in the morning. The plastic can be vented or slitted during the growing period (January, February) if the temperature increases beyond biological limit inside the tunnel during the peak day time. Generally 3 to 4 cm size vents are made on eastern side of the tunnels just below the top at a distance of 2.5 to 3 m after transplanting. Later on, with further increase of temperature the size of the vents can be increased or new vents may be added in-between two existing vents. Finally the plastic, used only once, is completely removed from the plants during March.

However, disposing of these waste plastic covers is a huge problem of such practice.

Fertigation: In low plastic tunnels the fertilizers are applied through drip irrigation system to provide a continuous supply of nutrients to the crop along with water. In the first month irrigation can be applied @ 4 m^3/ 1000m^2 area at an interval of 6 to 7 days. After making fertilizer solution of N:P:K (5:3:6), it is applied @ 80 to 100 ppm / m^3 of water. In second month the interval may be reduced to 4 days and the ppm increases to 120 to 150/m^3 of water. This will continue till flowering starts. After that the dose of fertilizer may be reduced to some extent and again increased during the fruit development stage.

If required, systematic insecticides can be applied through drip irrigation to control the infestation at early stages as, during that period, application from outside is a problem. Later it may be applied manually as per requirement.

In case of tunnels covered with shade-net or IP-net, mini-sprinkler or mist system of irrigation system may be adopted. Particularly for rising of seedling or planting materials this combination is very much useful.

4.6.2. Walk-in plastic tunnels

It is almost similar to the Quonset type of greenhouse. However, the only difference is that the heights restricted to man-height, i.e. about 6 to 7 feet at middle and no specific ventilation structure is provided. (See Fig. 21.b)

It is basically used to increase the temperature and humidity inside, particularly in dry cold climates. Due to this reason this walk-in plastic tunnels are called as 'hot-house'. In climates where the chilling winter season (lowest temperature close to zero) is dry, this low cost greenhouse has been proved to be more effective round the year for some specific crops (for example 'Kalimpong' in Darjeeling district of India).

Otherwise this type of low cost greenhouse can be used to extend the growing season of crop up to 2 to 3 months in a very cool climate where temperature goes well below subzero in the winter and reduces the growing period of crops.

Such walking tunnel can also be used in tropical or subtropical climate or for urban-farming purposes, if the 'roof ventilator' and 'side wall window' is provided in the structure. Thus, it should be designed accordingly (See the Fig. 23).

4.7. Drainage and Source of Irrigation

4.7.1. Drainage

Drainage system for greenhouse cultivation or even for open farm is an important factor. Two types of drainage system/planning are considered for quality farming practices, viz. 1) Drainage system for safe disposal of rainwater and excess irrigation water from surface of the farm. 2) Drainage system for removal of excess soil-water beyond the root-zone area. This system also prohibits the accumulation of excess water in the root-zone area.

In case of greenhouse, water is supplied to the crop in a very measured way and the rainwater is prevented to enter inside. Hence, drainage of rain-water is most important for greenhouse cultivation. The run-off created due to rain fallen on a greenhouse is almost 100 percent of the quantity of the rain/storm. As is the run-off area (roof of greenhouse) provided with sufficient slope and less friction, the amount of rainwater accumulated for immediate disposal is very high and requires a good drainage planning. Without proper drainage system for quick disposal of rainwater out of the farm area, the water may enter into the greenhouse through seepage or otherwise and destroy the crop or disturb the crop growth seriously.

Proper drainage system for greenhouse cultivation should consist of 1. Roof-water disposal system attached with greenhouse structure equipped with gutter rainwater pipes etc. 2. Open or underground drains are essential to collect and dispose the rainwater from the vicinity of greenhouse or to collect and carry this rainwater in a storage (rainwater harvesting) structure. 3. Main drainage channel leading towards safe disposal area. 4. Size and shape of beds/pots and the soil mixture inside should facilitate the drainage of excess soil-water below the root zone area.

In general, the drain pipes are connected to gutter and drains, situated outside the greenhouse. The drains are completely seepage and leakage proof which leads to the ultimate drainage channel or water-harvesting structure. The slope and size of the gutter drain pipes and surface drains should be designed in such a way that it can quickly dispose of the peak run-off water.

Apart from this, the open area existing in between and surroundings of the greenhouse should be developed in such a way that the rain water is disposed far away from the greenhouse. For this matter earthen channel with an earthen ridge, along the wall of the greenhouse, should be constructed. This will prevent the rainwater to accumulate in the open area and seep inside the greenhouse.

4.7.2. Source of Irrigation Water

Source of irrigation water is an important aspect of greenhouse planning. It includes 1) Availability of irrigation water, 2) Quality of irrigation water, and 3) Cost of irrigation water.

For greenhouse cultivation, a relatively expensive technology, the collection of information about availability of suitable irrigation water near the site is a prima-facie requirement. Thus, before purchasing a site for greenhouse project the available water source should be tested for quality and quantity. Only quality water is permitted to use for greenhouse for irrigation purpose. And as a general rule 20 liter water is required to irrigate one square meter of greenhouse area. However, the required quality and quantity of irrigation water depends to some extent on methods of irrigation.

4.7.2.1. Quality of water

Chemical content of water is very important. High salt content is a very common problem. Sodium and Boron content sometimes create problem. Removal of salt is expensive and hence, not commercially feasible for greenhouse cultivation. The pH of water should be around normal (6 to 7). Higher concentration of salts raise the pH and adjustment of the pH level of the water is also expensive.

Table 15: The class and corresponding quality of irrigation water

Class of Water	Electrical Conductivity (mho/cm x 10^{-5} at 250 C)	Total Dissolvedsalts (ppm)	Sodium (% of total salts)	Boron (ppm)
Excellent	< 25	< 175	< 20	< 0.33
Good	25 - 75	175-525	20-40	0.33-0.67
Permissible	75 - 200	525-1400	40-60	0.67-1.00
Doubtful	200 - 300	1400-2100	60-80	1.00 -1.25
Unsuitable	> 300	> 2100	> 80	> 1.25

Apart from these yardsticks few other chemicals in higher concentration, like Chloride, Fluoride, Arsenic and sometime Heavy Metals makes the water non-usable. This occurs due to urbanization or other reasons.

Furthermore, different types of irrigation system also require specific standard of irrigation water. Quality of irrigation water essential for drip irrigation system is discussed in chapter 7.5.3.2.(a).

4.7.2.2. Source of quality irrigation water

Rainwater is the best quality water for irrigation purposes. Next is underground water pumped out from deep aquifer.

Rainwater

Many regions have a high water surplus during rainy season. In greenhouses the entire amount of rainwater is surplus i.e. the amount running-off from the roofs of the greenhouses through drainpipes. This water can be collected and used for irrigation purposes. A sufficiently large storage structure is required to be built depending upon the quantum of rain, its distribution and intensity. Now, the crop-water requirement can be linked with this amount of stored water to assess the requirement of irrigation water for greenhouse cultivation.

To design the storage structure for rainwater harvesting for creating a resource for quality water, the following points have to be considered.

- Estimate the greenhouse area and the open land available to create catchments area for the storage structure.
- Amount of rainfall of a single rain/storm considering the intensity and distribution.
- Amount of irrigation water required.
- Maintenance and cleaning of storage basin.

Calculation of storage volume In principle, the daily variations and the frequency of rainfall have to be taken into consideration for accurate calculation of storage volume for rainwater. But it is almost impossible to collect these values. Hence, the mean monthly rainfall value is approximately considered. For a greenhouse this can be calculated in the following way.

$$CV = P \times f_c \times A_g$$

Where: CV = Calculated Volume (liter/month)

P = Precipitation (liter of rainwater / sq. meter / month)

f_c - Collecting factor (ratio of run-off to rainfall - for greenhouse it is 0.9)

A = Area of floor of greenhouse (sq. meter)

Monthly precipitation available for storing (StPm) is:

StPm = CV / A - WR / Ag (liter/sq meter/month)

Where WR = water requirement of crop (in liter) per month.

The value of StPm may be positive, may be negative. Positive means the rainwater available for storage is excess and sufficient enough to irrigate the crop. On the other hand, negative means the water is deficient to that value and arrangement for other source of irrigation water is required.

For final estimation of storage volume the Yearly Storageable Precipitation (StPy) is to be calculated from the StPm figures, calculated separately for every month. Then addition of the positive figures (StPy) and negative or deficit figures (DeFy) are done separately. The yearly storage balance (StBy) is then determined by subtracting the DeFy from StPy.

The yearly StBy may also be positive or negative. In the first case the storage volume (Vst) is predominantly guided by the WR. In the second case, the higher value in between StPy and maximum monthly CV is to be considered to determine the required storage volume **(Vst)**. However, for estimation of Vst the coefficient of variation for precipitation (V_c) is to be considered.

Example Calculation of storage volume for two different cases, on the basis of given StPm figures.

	Jan	Feb	Mar	Apr	May	June	Jul	Aug	Sept	Oct	Nov	Dec
C-1	49.1	34.2	29.3	6.9	-15.5	-56.0	-86.3	-65.5	-6.8	75.4	98.7	86.6
C-1	32.3	21.1	5.0	-33.7	-64.3	-85.7	-104.7	-96.4	-75.2	-9.5	43.1	52.5

For case-1: StPy = 380.2, DeFy = 230.1, so the StBy = 380.2 - 230.1 = 150.1
         Hence the Vst = 276 liter/m2 of greenhouse [DeFy x (1 + V_c)]

For case-2: StPy = 154.0, DeFy = - 469.5, so the StBy = 154 - 469.5 = - 315.5
         Here the StPy > maxCV/Ag (104.7 in July)

Hence the Vst = 158 liter/m^2 of greenhouse [StPy X $(1 + V_c)$]

If the StPy < max CV/A$_g$ then Vst = maxCV/Ag x $(1+V_c)$

Storage structure: In greenhouse projects the rainwater harvesting structures may be of following type

- **Earthen structure**: Clay type subsoil with sufficient watertight bottom is prerequisite for this structure. It is cheap but poses problems in respect of maintenance, supply of clean water and seepage loss.

- **Earthen structure with plastic lining**: In light textured subsoil, pond like storage structure with lining of black PE-film (0.2 to 2 mm thick) or PVC-film (0.4 to 2mm. thick) is useful. In comparison to earthen structure the average cost/rn^3 with PE- film and PVC-film is almost double and triple respectively.

- **Concrete or brick structure**: It is expensive but durable with low maintenance cost. However, it should be designed properly, particularly when it is constructed aboveground. The cement lining should be done carefully to make the structure completely waterproof. The cost of such structure is about 10 to 15 times to the earthen structure.

Arrangement to cover the surface of all kinds of reservoirs should be made in order to prevent excessive evaporation loss.

Groundwater

As other source of irrigation water, groundwater may be tapped through digging of open (large diameter) wells or drilling of bore wells or sinking of tube wells. The water is lifted from these wells by indigenous ways or by using pump sets. Depending upon the discharge capacity in relation to the irrigation water requirement, all these types of groundwater sources can be used for greenhouse project. However, the choice between 'open well' and 'tube well depends upon (a) availability of space, (b) status of underground aquifers with corresponding quality of water, (c) topography, (d) condition of underlying rocks, (e) rainfall, (f) status of recharge with seasonal fluctuation, (g) availability of water lifting system and finally the (h) cost with economics.

After making necessary choice for the type of well, the design in respect of diameter, depth and lining/casing with filter is required to be made. After that the method of construction is selected.

For high discharge wells the groundwater may be used directly for irrigation purposes. For low discharge wells and as a common practice, the groundwater may be stored in a storage tank and subsequently used for irrigation purposes.

4.7.2.3 Materials required for drip irrigation system necessary for greenhouse

High quality drip irrigation system (reputed make accessories & pipes):

1. ½ to 1 hp Pump,
2. Sand & disc filter,
3. Backwash assembly,
4. Ventury,
5. Inline/external drippers,
6. Main, sub-main, and lateral (16mm), pipes,
7. Gate/pressure release/ball valves
8. Connecter, etc.

4.8. Estimates and Structures of Different Types of Greenhouses for Average Indian Conditions

4.8.1. 'A' type broken roof greenhouse structure with 50mm pillar and pipe roof

Area: 30.00 x 6.00 = 180.00 m²

Pole spacing: along width =3m, along length = 4m

Ventilation: 100 % of floor area and covered with IP-net.

S.No	Item	Quantity	Rate (Rs.)	Amt. (Rs.)
1.	S, F, Fixing G.I.pipe.			
	(a) 50mmdia (m)	100.00 m		
	(b) 25mmdia(m)	380.00 m		
2.	Earth work in excavation & foundation trenches....etc.	25.00 m³		
3.	Cement concrete 1:3:6.	5.00 m³		
4.	S, f, & fixing of door chamber	5.04 m²		
5.	Supplying & laying of PE-film	390.00 m²		
6.	S, F,& fixing of IP-net etc.	180.00 m²		
7.	Misc. items clip, rope/wire etc.	L.S.		
Sub-total				
Add 5% contingency				
Approximate Total				
Approximate rate per square meter = Rs. 700.00				1,26,000.00

4.8.2. 'A' type broken roof greenhouse with 40mm pillar (pipe) & MS-frame roof

Area : 30.00 x 6.00 = 180.00 sqm

Pole spacing : along width = 3m, along length = 4m

Ventilation : 100 % of floor area and covered with IP-net.

Sl. No	Item	Quantity	Rate (Rs.)	Amt. (Rs.)
1.	S, f,& fixing G.l. pipe.			
	(a) 40mmdia.(m)	100.00m		
	(b) 25mmdia.(m)	250.00m		
2.	S,F & fixing of MS angle	0.238 MT		
3.	Earth work in excavation in foundation trenches.	25.00m²		
4.	Cement concrete 1:3:6	5.00 m³		
5.	Providing door	5.00 m²		
6.	S & Laying of PE-film	390.00 m²		
7.	S,f &fixing net...etc.	180.00 m²		
8.	Misc. items like clips, ropes etc	L.S.		
Sub Total				
Add 5% contingency.				
Approximate rate per sqm = Rs. 600.00.		Total		1,08,000.00

4.8.3. Estimate for pipe frame 'Quonset'/walk-in tunnel Greenhouse (4m width)

Area: 4.00 x 24.00 = 96.00 m² (Fig. 24)

Ventilation: 64 % of floor area and covered with IP-net

Sl.No	Item	Quantity	Rate (Rs.)	Amt. (Rs.)
1.	S, f& fixing G.l. pipe.			
	(a) 25mm dia (m)	150.00 m		
	(b) 38mmdia (m)	30.00 m		
2.	S, f & fixing G.l sheet	0.864 m³		
3.	M.S.structural work.	8.70 kg		
4.	S, F,&fixing nut,bolts etc.	20.00 kg		
5.	Cement concrete 1:3:6.	1.00 m³		
6.	S F &laying PE-film	154.00m²		
7.	Supplying&laying I P-net.	62.00m²		
8.	Wood work.	0.15 m³		
9.	Misc.	L. S.		
Sub-total				
Add 5% contingency.				
Approximate rate per sqm = Rs 500.00/sqm.		Total		48,000.00

4.8.4. Estimate for pipe frame 'Quonset' /tunnel type Greenhouse (6m width)

Area : 6.00 x 24.00 = 144.00 m^2

Ventilation : 64 % of floor area and covered with IP-net

Sl.No	Item	Quantity	Rate (Rs.)	Amt. (Rs.)
1.	S, f, & fixing G.l. pipe (a) 25mm dia (m)	252.00m		
	(b) 15mm dia (m)	120.00m		
2.	M.S stactural work.	30.00kg		
3.	G.l sheet.	1.08 m^2.		
4.	C. C. 1:3:6.	2.23m^3.		
5.	Polythene sheet	350.00m^2		
6.	**I. P. net.**	140.00m^2.		
7.	Wood work	0.20m^3		
8.	U. clamp	210.00 nos.		
9.	G.l. nipple	21.00 nos.		
10.	Misc.	L.S.		
Sub-total				
Add 5% contingency				
Approximate rate per sqm = 550.00/sqm.				79,200.00

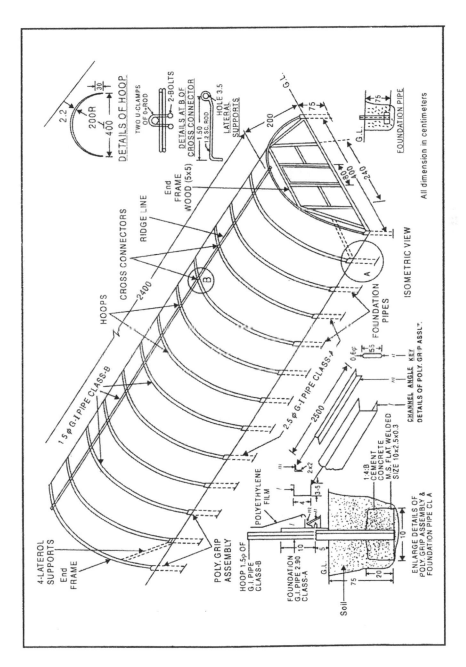

Fig. 31. Details structural design of pipe frame Quonset type greenhouse

4.8.5. Estimate for flat-roof shade/net house using wire at roof (Fig. 25)

Area : 16.00 x 16.00 =256.00 sq m.

Pole spacing: 4m x 4m

Sl.No	Item	Quantity	Rate (Rs.)	Amt. (Rs.)
1.	S, F & fixing 50mm dia G.l pipe,	150.00 m.		
2.	Earth work in excavation in foundation, trenches etc.	10.00 m^2		
3.	Cement concrete 1:3:6.	7.50 m^2		
4.	S, fitting & fixing 7 ply galvanized wire 19 gauge.etc.	500.00 m		
5.	S,fitting & fixing straining bolt.	25.00 nos		
6.	S, fitting & fixing net....etc.	475.00 m^2		
Sub-total				
Add 5% contingency.				
Approximate rate Per sqm = Rs 350.00				
Total				89,600.00

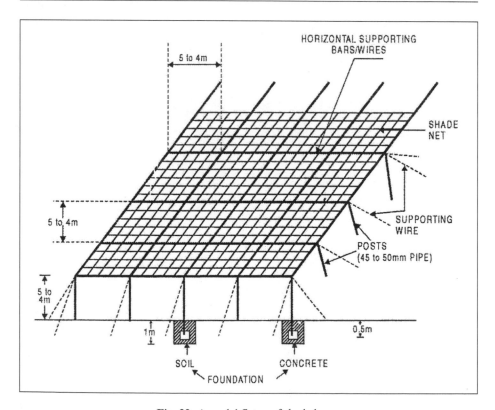

Fig. 32: A model flat-roof shade house

4.8.6. Construction of wooden structure 'A' type greenhouse

Area: 6 m x 24 m = 144 m^2

As per the specification prepared by IPCL, INDIA (figure - 26)

Sl. No	Item	Quantity	Rate (2005)	Amt. (Rs.)
1.	5x5 cm wood (roof) -3m each	25(75 m)		
2.	5x5 cm wood (roof) - 4m „	50 (200 m)		
3.	5x2.5 cm wood (purlins) - 2.5 m „	52 (130 m)		
4.	7 - 10 cm dia. Pillars - 2.5 m „	25 (87.5 m)		
5.	7 - 10 cm dia. mid. pillars- 3.5 m „	6 (24 m)		
6.	Small (15 cm) long pole	104 (16 m)		
7.	Support for PE-film- 4mm G I wire	410 m (40 kg)		
8.	3mm thick MS strip for joints	0.375 m^2		
9.	UV PE-film - 200 micron	320 m^2 (60 kg)		
10.	Antitermite coating – Bitumen	10 liter		
11.	Black LDPE strip for jointing	3 kg (20 m^2)		
12.	Foundation - 1:3:6 CC	2.21 m^3		
Sub-total				
Add contingency @ 10%				
Approximate rate Per sqm = Rs. 600/m^2		Total		86,400.00

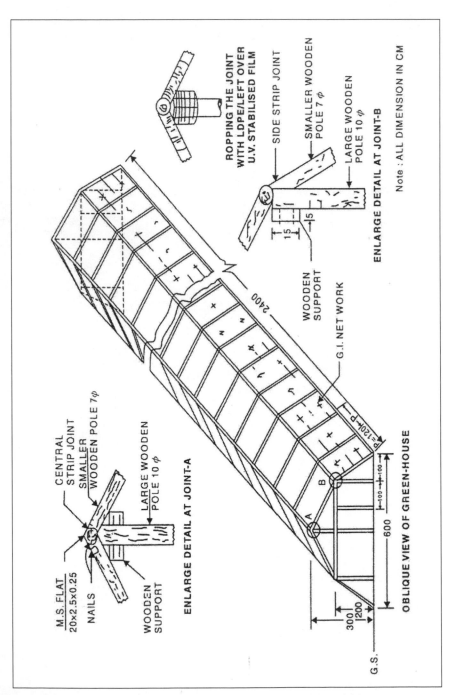

Fig. 33: Drawing of Wooden structured greenhouse, by IPCL.

4.8.7a. Crude Bamboo frame greenhouse

Area: 6m x 18m = 108 m^2

The design is developed by Assam Agriculture University, Jorhat, India (figure-27)

Sl.No	Item	Quantity	Rate (2005)	Amt. (Rs.)
1.	Large dia. Bamboo (2.25m)- roof	26 (58.5m)	17.00/m	995.00
2.	Small dia. Bamboo (6 m)- purlin	9 (54 m)	13.00 /m	702.00
3.	Bamboo for side & foundation (3m)	26 (78 m)	17.00 /m	1326.00
4.	Bamboo pillar for center (3 m)	13 (39 m)	17.00 /m	663.00
5.	Bamboo pillar for side (2 m)	26 (52 m)	17.00 /m	884.00
6.	Large dia bamboo for side (6 m)	18 (108 m)	17.00 /m	1836.00
7.	Small dia. Bamboo for end frame	- (47 m)	13.00 /m	711.00
8.	Door	- (18 m)	13.00 /m	229.00
9.	UV PE-film – 200 micron	256 m^2 (48 kg)	57.00 /m^2	14592.00
11.	Anti-termite coating – Bitumen	10 liter	13.00 /lit	130.00
12.	Foundation – 1:3:6 CC	2.5 m^3	1818.00 /m^3	4545.00
Sub-total				26,613.00
Add 10% contingency				2,661.00
Approximate rate Per sqm = Rs. 271/m^2		Total		Rs. 29,173.00

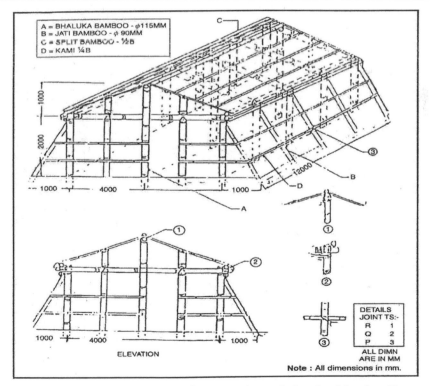

Fig. 34: Structure of Bamboo-frame greenhouse designed and developed by
Assam Agriculture University, Jorhat, India

4.8.7b. Innovative and comparable Gutter-Connected Bamboo greenhouse

The model below is innovated by the Author in 2011 to compete with pipe frame greenhouse in respect of efficacy. It is less clumsy, durable, and expandable to any proportionate size. (For detail please see chapter 5)

Area: 10m x 20m = 200m^2

Specification: Bay size-5m, Post to post-4m, Gutter height- 3m, Ridge height-4.5m, Side window-2m

Cost: Approximate Total Cost Rs. 90,000.00 @ Rs. 450/sqm.

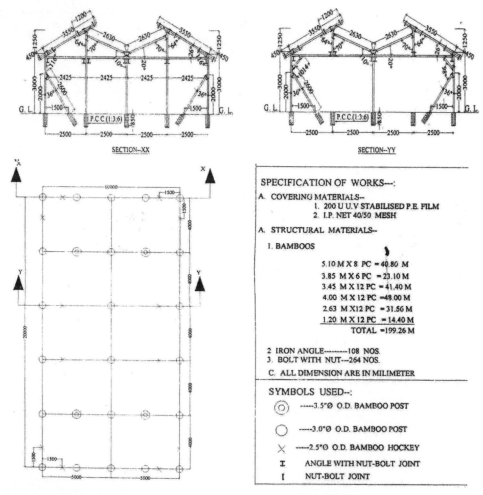

Fig. 35: Model of 200m^2 'bamboo greenhouse' innovated by the author

4.8.8 Estimate of structure of gutter-connected 'saw-tooth' greenhouse (Fig. 28.A)

Area- 36.00 X 24.00 = 864.00 sqm

Specification- Bay width = 6 m. Pole to pole = 4m, Gutter height 4.5m, Top height-6m

Sr.No	Item	Quantity	Rate	Amt. (Rs.)
1.	S, f,& fixing G.l. pipe.			
	(a) 50mmdia.(m)	744.00m		
	(b) 32mmdia (m)	432.00m		
	(b) 25mmdia.(m)	528.00m		
2.	S,F & fixing of MS angle	0.238 MT		
3.	MS flat foundation grip	20kg		
4.	Cement concrete 1:3:6	5.00 m^3		
5.	Door & ventilators(wood)	1.5 m^3		
6.	S & Laying of PE-film	1452.00 m^2		
7.	S,f &fixing net...etc.	518.00 m^2		
8.	Gutter(22x19cm)16gauge	62 m^2		
9.	Misc. items like clips ropes etc			
Sub-total				
Add contingency, Taxes, transport etc.				
Approximate rate per sqm = Rs. 950 / sqm				
	Total			8,20,800.00

4.8.9 Estimate of structure of gutter-connected 'Ridge & Furrow' (aerodynamic) greenhouse (Fig. 28.B)

Area- 36m X 28m = 1008.00 sqm

Specification- Bay width = 8 m. Pole to pole = 4m, Gutter height 4m, Top height-6 m.

Sl.No.	Item	Quantity	Amount (Rs)

1. Greenhouse Structural as per below
1. Column(67 or 60mmØ)
2. Arch/purlin/bottom truss/foundation bush (38/ 42mmØ)
3. Different 'Tanas'/connectors etc (32 & 25mmØ)
4. Channel & spring(12 gauge)
5. Clumps (16 gauge full & half,-1", 1.25", 1.5" ,2/2.5")
6. Gutter (24 gauge)
7. Grouting with CC (1.5-2 cft/grout)
8. Other items (wire, screw, ring, angle, etc & nut-bolts) made up of GI,

Contd.

2. Covering materials
1. Multilayer 200µ UV stabilized PE-film,
2. Removable 50% shade-net, and
3. 40/50 mesh UV Insect-proof (IP) net.
    Sub-Total
    VAT @ 5%
4. Labour charges (for Installation at site)
5. Material processing charges
6. Technical assistance/supervision charges
7. Transport (taken an average distance)

Approximate rate per sqm = Rs. 850.00   Total   8,56,800.00

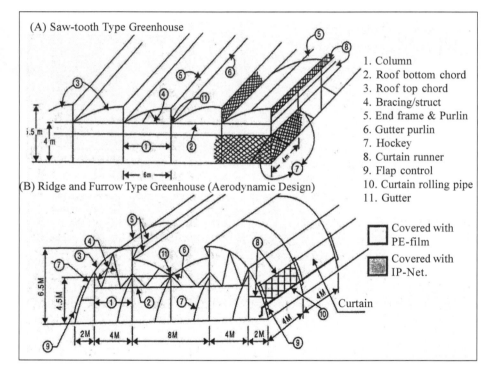

1. Column
2. Roof bottom chord
3. Roof top chord
4. Bracing/struct
5. End frame & Purlin
6. Gutter purlin
7. Hockey
8. Curtain runner
9. Flap control
10. Curtain rolling pipe
11. Gutter

Fig. 36. Structural Details of Gutter-Connected Greenhouse
a) Saw-tooth type and b) Ridge & furrow type.

5

Bamboo Greenhouse Technology
A Multi Stage Technological Innovation

5.1. Introduction

Back drop: Since the concept of greenhouse technology was introduced, metal and wooden structures were tried in different parts of the world. When greenhouse was brought in the tropical and sub-tropical areas, people tried bamboo as construction material for greenhouse instead of iron or wood. However, due to the non-standard character of bamboo, the standardization of design of bamboo greenhouse, as a model, was not possible. Actually all such designs were very much clumsy and not at all comparable to the well established models of greenhouse having metal or wood frame.

Hence it is an urgent call to establish a standard design of bamboo greenhouse that can efficiently compete with the other common & commercial greenhouses made-up of GI-pipes.

Innovation: This greenhouse is structured by matured, cured & treated bamboo instead of usual structural elements (GI/MS/wood). For that matter (1) innovated design and construction methods and (2) a new curing method has been generated. It is very tough to standardize the design of bamboo greenhouses, specifically due to the non-standardized nature of bamboo. Unlike GI/MS pipe etc, bamboos do not have the same dimensions and uniformity in size & shape (diameter/ width/thickness etc). Thus, in general, bamboo greenhouses are low in height and are of naturally ventilated type. However, there is still huge scope to work on design of bamboo greenhouse for different climatic situation to avail the benefits of its natural advantages and low cost.

What is the innovative Bamboo greenhouse?

- It is a package of technology innovated to construct an eco-friendly, energy-efficient, low-cost naturally ventilated greenhouse suitable for Humid climate.

- It is comparable to common greenhouse made up of pipe in respect of strength, cleanliness, durability (10yrs), and usefulness with a little more maintenance.
- It is similarly effective to perform precision farming under it.
- It is simple, scientific, affordable, and can accommodate moderate climate control mechanisms, but it is difficult to execute.

5.2. Bamboo: A Plant Material – the Unique Structural Element

Botany & Growth: It is a group of perennial evergreen plant in the true-grass family *Poaceae*, subfamily *Bambusoideae*, tribe *Bambuseae*. Bamboos are some of the fastest growing plants in the earth due to its unique rhizome-dependent growth system.

Unlike other trees/plants, the stems of bamboo, or culms, emerges from ground almost at their full diameter and completes its total height within one season (3-4 months) without any branching and leafing. It was reported that the maximum growth rates of bamboo in tropical areas is about 100 cm (39 in) in 24 hours. However, the growth rate is dependent on local soil and climatic conditions as well as species. In temperate area the growth rate is much less, about 3-10 cm (1-4 inches) per day.

After attaining the full height, the branches extend from the nodes and leafing in the branches occurs. In the next year, the pulpy wall of each bamboo/culm slowly hardens due to the lignifying process. During the third year, the culm completes its ongoing hardening process. At the same time the silicification of outer tube-wall occurs. After 3rd year it is considered as fully mature culm/ bamboo as it has completed both the Lignifying and Silicification processes that give it hardness and strength, particularly the tensile strength. Then only bamboo or culm is harvested or cut down for necessary use after necessary curing.

If the bamboo is harvested or cut-down before its maturity, ie. before 3 years, it will not last long and also not provide the potential hardness and strength, despite endowing this immature bamboo with all sorts of curing & treatment procedures.

Life of Bamboo: If the bamboos or culms are not harvested after attaining full maturity (3-4 yrs of age) then over the next 2–5 years (depending on species & climate), fungus and mold will begin to form on the outside of the culm, eventually penetrating and destroying the culm. Thus, at around 5–8 years of age (depending on species and climate), the fungal and mold growth will cause the culm to collapse and decay completely.

Matured Bamboo after proper curing and treatment is a wonderful product of nature in respect of size, lightness, strength, durability, and purposefulness, particularly when compared with other such plant materials.

Av. Approximate Mechanical Parameters of Bamboo (in kN/ cm²)

Elastic modulus - 2000, Compressive strength – 8.5, Tensile strength- 15 to 38, Bending strength - 8 to 27, and Shearing strength - 2. [Taken from Deutsche Bauzeitung (DB) 9/97]

However it widely varies on bamboo spp. & age, soil & climate, curing/seasoning, and also varies along the length of a single bamboo.

Chemical composition of Bamboo that is responsible for its character:

- Apart from water, bamboos consist of hemicelluloses, pentosans, lignin, etc with different combination of their proportion and provide the necessary strength. These figures vary widely with species, variety, age, soil, and climate.

- In addition, it contains ligno-cellulosic materials, and starch, which have very low resistance to biological degrading agents.

- The lignin present in bamboos is unique, and undergoes changes during the growth of the culms. Lignin is the binding material for the fibrous cellulose material and resist fungal & pest invasion.

- Bamboo is also known to be rich in silica (0.5 to 4%), but the entire silica is located in the outer layer (1 mm) that protects bamboo from any physical damage and biological infestation.

- It also has minor amounts of waxes, resins and tannins, but none of these have enough toxicity to improve its natural durability.

Thus, we can conclude that without artificial seasoning/curing, it will not last long.

5.3. Curing of Matured Bamboo

What is curing/seasoning: It is the process that reduces the vulnerability of fresh harvested bamboo towards infestation of any natural biological agents and increases its strength by manipulating its physical and chemical properties.

Reduction of water or drying after drenching out the sugar are to be done in a selective form, this is the basic concept and method of curing.

History: Two ancient basic methods were used by our predecessors. First method is prolonged dipping of bamboo in water followed by proper sun-drying. Storing bamboo in water or "leaching bamboo" followed by drying is a traditional bamboo preservation method, used by indigenous communities and farmers of several Asian and Latin American regions. This will followed by a normal-drying process. In Latin America it has been the tradition to transport bamboo from the mountain and jungle areas towards the urban centres by means of bamboo rafts. This act of dipping in water was actually used to drench out the sugar and to some extent carbohydrate present in the matured bamboo, thus after drying it increased the relative percentage of lignin in the mass of bamboo. The **Second** method is the moderate & controlled roasting of bamboo by exposing it to restricted fire. This will reduce the water present in the matured bamboo to make it dry, light, and strong. This controlled firing will continue until the resin and other chemical rises to the surface and/or convert into other biological/ organic compounds. According to the method of Japanese craftsmen you can build a charcoal fire-pit to cure your bamboo with this heating strategy.

Once your curing is complete the bamboo will generally change from green to a yellow/tan colour. Cured bamboo is very light but extremely strong. This makes it great for a multitude of uses.

Modern methods: (a) dipping the bamboo for hours in plain hot water or in a closed chamber of pressurized hot water (along with chemicals) followed by drying, (b) few sap-displacement methods are also used by means of gravity or by pressure (pump)/ex-osmosis, and (c) treat high pressure water vapour (about 100°C) in a closed chamber filled with large number of sized whole bamboo. It is a method innovated by the author to cure truck-loads of bamboo within about 6 hours only. It is followed by immediate dipping in chemical solution and then drying.

All these curing methods are used to drench out the water along with sugar, resin and to some extent carbohydrate present in the matured bamboo.

Problem of these methods: However, except the innovative high pressure water vapour treatment, all other methods are not useful to cure large numbers of whole bamboos at a time and within a short period. However any of these methods can be adopted with proper planning.

Specific requirement of curing for greenhouse structure: Curing of bamboo is an important aspect to construct a bamboo greenhouse. For that purpose two aspects should be considered carefully; (1) first, straight and matured (3 to 4 yrs of age) bamboo/culm should be collected, (2) Second, the sized bamboos should be cured in bulk, in a short period.

In this context, for curing, the innovative 'pressurised water-vapour' method was successfully tried where large numbers of whole bamboo are exposed to high pressure hot (about 100°C) water vapour in a closed chamber to cure it within about 6 hours (Photo:1). The pressurized hot water vapour is supplied from a boiler through pipeline and released by perforated pipes placed in the said chamber (applied for patent in India in 2011).

Photo- 1: Curing & treatment of whole bamboo (in bulk) innovated by the author.

5.4. Treatment of Cured Bamboo

After proper curing, which gives it the strength and necessary physical character of a structural material, the treatment is required to protect bamboo for future attack of different pest and pathogens. Now-a-days few modern curing techniques also do the treatment process simultaneously, like the exposure of bamboo to high pressure water along with the borax and boric acid solution in a water tight chamber.

Many established chemical methods of treatment are available to protect bamboo against insect pests and fungal attack. However, for greenhouse construction, made with poly-ethylene film, all such chemicals may not be suitable. Some of these chemicals may damage the film when it comes in contact with the treated bamboo. The treatment of bamboo with boric-acid and borax solution, in a suitable proportion has been found effective. The clean and treated bamboo should be cut into required pieces and then dip into the said solution for minimum six hours. Through this process the boron salt, *disodium-octaborate*, is precipitated on bamboo, in concentration of 5-10%, and protect it from both fungus and pest attack. As this *disodium-octaborate* is water soluble in nature, this type of treated bamboo should not be exposed to water or excessive moisture.

5.5. Drying and Other Processing of Cured and Treated Bamboo

To add more effectiveness to such curing and treatment methods, a further processing like drying, painting/coating/covering, etc are to be done in a selective and meticulous manner. After properly drying the treated bamboos can be covered with any suitable water resistance protective layer (with coal-tar, bitumen, resin, oil) or paint.

Drying is an important factor as because it will invite shrinkage (10-16% in diameter and 15-175 in thickness) and crack to the round/non-split bamboo if not the method is designed properly. Quality and age of bamboo is the first criteria to avoid the above mentioned problems. Thus bamboo with 3-4 years of age and about 1 inch wall thickness can provide proper result of drying of round-bamboo

For round bamboo, drying should be done slowly and will avoid direct sunlight when 'sun-drying' method will be followed. Rapid change of moisture content of round bamboo may causes cracks in internodes portion and reduce the strength.

While drying, the round bamboo should be stacked vertically under partial shade. The cured and treated round bamboo needs a minimum 15 days of such drying. However, the drying process will continue thereafter with a very slow rate and continue for months together.

Boric-acid & borax solution

Photo 2: Treatment of bamboo

Jacketing of posts after painting of bitumen

Sizing & drying of bamboo

Photo 3: Preparation of posts for Direct grouting- painting of tar and jacketing

5.6. Problems Related to Structural Design of the Bamboo Greenhouse

Problems: There are many problems of using bamboo to design and construct a standard greenhouse model, which are as follows:

1. The non-uniform cross-section corresponding to the length of bamboo;
2. Difficult to join two bamboo pieces properly and strongly;
3. Selection and availability of suitable variety of bamboo;
4. Lack of Proper technology of curing and treatment of mature bamboo in bulk;
5. Lack of standard grouting technology of bamboo posts;
6. Lack of experienced craftsmanship to adopt clean and standard design of a bamboo greenhouse model. No standardization of craftsmanship is possible

Ways to tackle the above mentioned problems: The probable solutions developed to overcome the above mentioned problems are stated below point-wise.

1. Non-uniformity of Bamboo: As bamboo is a plant-based material, it is not possible to do anything in respect of non-uniformity of bamboos, and it is always be considered and used as non-standard material. However proper selection of variety, individual piece/length of bamboo component, and craftsmanship are the ways & means to reduce this problem to a great extent.

2. Problem of joining of Bamboo: Different methods are tried and are still trying to solve the problem of joining of bamboos at different points of a structure. Using different nail/bolt and binder are commonly tried since past. However, in 2011 a new method to join the bamboos using MS-angle and GI bolt was developed (pic: 7) which is found to be very much effective in comparison (it is a part of patent applied in 2011). This method of joining was already applied in construction of about 1700 no small greenhouses at Sikkim, India at farmers' field.

3. Suitable variety of bamboo: The availability of suitable variety of bamboo is dependent on area or locality. However, proper variety of bamboo with due treatment & curing can be transported to distant places of execution. The selection of variety should be corroborated with the need of different structural elements of bamboo greenhouse. The diameter, corresponding wall thickness and inter-node spacing of bamboo are the three vital criteria of selection of bamboo variety. For example, it is always better to select two different varieties for using it as bamboo post/pillar and bamboo praline according to the requirements of these elements. In general 3-4 inch diameter having closer internodes and thick wall bamboo variety is suitable for post and hockey, and 2.5-3 inch average OD with long internodes and moderately thick wall bamboo variety is suitable for construction of truss elements. However, among many suitable bamboo variety the most suitable species of bamboo are 1) Bambusa Balcoa, (balku/ baro - bans), 2) Dendrocalamus giganteus (Bhalo Bans), 3) Bambusa nutans, (Makla), 4) Dendrocalamus hamiltonii (Pecha/ Tama), of which the 1st two are used for posts and the last two are used for purlin and 'tana' i.e. roof structure.

4. Curing and treatment of bamboo: discussed earlier.

5. Grouting of bamboo: Grouting of bamboo is a real problem to construct bamboo greenhouse. Contact of bamboo with soil increases the possibility of decaying the bamboo. Hence, something should be done to grout the bamboo post and hockey in the ground that restrict the grouted bamboo to come in contact with soil. According to my knowledge/practice and available information two standard methods has been innovated and considered useful to solve this problem.

First method: In this method the treated bamboo posts/pillars is first treated with bitumen and then jacketed with water-proof plastic jacket. Here twice the length of grouted portion (3-4 ft) from bottom will be painted with diluted bitumen and inserted the portion into a leak-proof plastic jacket (3-4 ft long) made up with 150 micron thick UV-resistant PE-film and the top portion of the jacket should be sealed properly around bamboo. Then half of this treated portion of bamboo post will be inserted in foundation pit perpendicularly and packed with cement concrete (CC). The half of the jacketed portion which remain above ground will protect the lower part of the bamboo from soil splashed due to rain (photo: 8).

Second method: A one inch diameter GI-pipe or any suitable section (angle/ rail/etc) will be grouted in the ground with CC, projecting a portion (2-3 ft) above the ground. Alternatively, a pre-cast readymade CC grout having reinforcement with 1inch diameter GI pipe or any suitable section (angle/rail/ etc) of which 2-3 ft is projected out from the top can be produced in factory (photos: 8). Then it can be transported to the site for making the foundation of the bamboo greenhouse (applied for patent in 2011).

The pre-cast unit will be placed in the ground leaving out the projected portion of the pipe above the ground. Then it will be covered and compacted with soil properly. Water will be applied to compact the soil further. After completion of grouting, the lower end of bamboo posts (1-2 ft short of the height) having a min 1 ft hollow inter-node portion will be inserted into the pipe of the grout. It shall be fixed with suitable cement-resin based glue and fitted to a clump (photo s: 8). Alternatively the projected portion of angle/rails will be bolted with the lower

Photo 4: Model of stand-alone open window bamboo greenhouse

Photo 5: Inside view of bamboo greenhouse

Photos 6: Inside view   **Photos 7:** Different types of joints

Direct grouting   Pre-cast grout & grouting with
                              pre-cast grouts
Photos 8: Grouting technology of Bamboo (2 types)

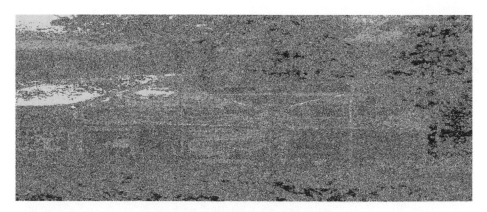

Gutter connected bamboo greenhouse

Gutter-connected 600 sqm Bamboo Greenhouse

portion of the bamboo posts leaving a gap above the ground. Here the bamboo has no connection with ground and even soil splashed by rain will not reach the bamboo post.

5.7. Design of Bamboo Greenhouse

Problem in design: Considering the load distribution, particularly the wind load and corresponding difficulties of joining bamboos, it is difficult to design the bamboo greenhouse neatly and cleanly. In most of the cases the necessary strength of the structure is provided by increasing the number of bamboo elements in different places of the structure, that makes the structure clumsy and poor to maintain.

Type of greenhouse: As most of the bamboo greenhouses are tried in tropical and sub-tropical areas, the necessary design should be of naturally ventilated type. Roof window and side window facility of such bamboo greenhouses was considered while designing. Considering the abovementioned problem, an innovative joining method has been developed (photos: 7) that helped to create the standard design of bamboo greenhouse. This design will be very similar to that of other common greenhouse in respect of cleanliness and strength to withstand the load. On that basis a basic structural model of such greenhouse was designed (photos: 4,5,&6), which can be replicated or modified to create such bamboo greenhouses suitable to a specific area, purpose, and climate (A kit of such designed structure is applied for patent in 2011).

Fixing of covering materials: The fixing of plastic film and/or net in such bamboo greenhouse is done by the same channel and zig- zag spring system as used in steel greenhouse. The only problem here is to fix the channel with bamboo frame.

Materials required: Apart from bamboo, as structural material, all other materials (Film, IP-net, shade-net, gutter, joining items, nut-bolts, etc) used for bamboo greenhouse are of the same quality used for steel-frame greenhouse. Now, depending upon the design, generally two types of bamboo are used to construct a bamboo greenhouse. For post and hockey the bamboo should be strong with thick (dd1 inch) wall, large OD (dd 4 inch), and medium (+/-1ft) size internodes. The suitable species of bamboo are 1) *Bambusa Balcoa,* (balku/ baro -bans) 2) *Dendrocalamus giganteus (Bhalo Bans)*. It is generally heavy and the cross section and OD tapering gradually towards the tip of the bamboo. Thus, it should be sized with proper planning to minimise wastage. For purlin and 'tana' the bamboo should be uniform along the length with long (+/-1.5ft) internodes, moderately thick wall (3/4th inch), and medium OD (2½-3 inch). The suitable species of bamboo for this purpose are 1) *Bambusa nutans,* (Makla) 2) *Dendrocalamus hamiltonii (Pecha/ Tama)*. It is commonly hard, glossy, pipe like bamboo having moderately thick wall. On an average the ratio of these two types required for a greenhouse is about 3 : 7

Fixing of different equipment and tools: This design is very much accommodative to the facility of different irrigation methods (drip, mini-sprinkler etc), exhaust systems (particularly where insect-proof nets are used in windows/ openings), air circulating fans, climate control techniques (like misting, fogging etc.), and usual drainage method (using gutter) used for multi-span bamboo greenhouse. However, high tech climate control systems like fan-pad cooling system, which require completely closed structure, may not be possible to incorporate in this design of bamboo greenhouse. However, with some modification the fan-pad cooling system can also be adopted with bamboo greenhouse comfortably.

5.8. Maintenance

Maintenance is the backbone in respect of life expectancy of bamboo structure. If a bamboo structure is maintenined properly and religiously then it can get a life more than 10 years. The maintenance of bamboo structure should be done in the following way.

1. **First maintenance** – It should be done 2 months after completion of the greenhouse. It is a simple work of tightening of all the nuts and bolts attached with bamboo frame. It is important because further drying up of the bamboos inside greenhouse may loosen the nuts and bolts. Consequently, this may loosen the structure as well as the covering material and start to create problem against high wind or storm. Thus it is simple but important.

2. **Yearly maintenance** – It is of two types. First, every year, before onset of the dry season all the nuts and bolts need to be tightened and simultaneously the condition of bamboos are to be inspected, particularly near every joint. If any kind of crack is developed, particularly along the line of bolt, then that part of bamboo should be strapped properly with any strong tape or rope or wire. At that time, if needed, necessary treatment of affected part of any bamboo piece can be done against any pest attack (with chloropyriphos + Kerosene, 4:1) or any attack of fungus or mold (fungicide/ oil/ paint). Second, every year before onset of monsoon the plastics and IP-nets are to be cleaned with long brush and pipe-water and, if necessary with mild chemicals. After that, the first shower will do its job to clean it further.

3. **Elaborate maintenance during change of PE-film**- After every 3-4 years when the covering film is required to be changed, a thorough and elaborate maintenance of bamboo greenhouse is to be done properly. It is commonly done before the dry spell of the year. At that time, each and every pieces of bamboo is to be inspected carefully and necessary maintenance work should be done against all the affected pieces. Any suitable protective measures like chemical treatment, oil (used engine oil) application, painting, etc may be done on the relatively susceptible bamboos of the structure. If any piece of bamboo is found to be very weak and potentially not to last for the next three years, then that piece should be unscrewed from the structure and replaced by strong cured and treated bamboo piece. In case of direct grouting it may not be possible, and in that case a supporting bamboo post should be bolted with the affected post.

5.9. Advantage and Disadvantage of this Structure Over Metal Structure

Advantage

- The cost of bamboo structure is almost 1/5th of metal structure, and the total cost of bamboo greenhouse is about ½ of the steel-frame greenhouse.
- The low initial cost with similar production capacity makes this bamboo greenhouse the most financially viable option and also makes it affordable to small farmers.
- As the height of bamboo greenhouse structure is shorter (3m gutter-height and 4.505m ridge height) the overall wind load is less in comparison.
- As the gable width is less (5m) in bamboo greenhouse, the length of roof window per unit area is higher than metal structure greenhouse, thus have more efficient roof window.
- As the thermal conductivity is low in bamboo, it do not absorb and release inside heat, thus help to reduce the inside temperature up to 2°C in comparison.
- As the side wall is straight or slanting (in high rainfall areas) and covered with IP-net (except apron), the air intake through side-wall ventilation is much more and increase the efficiency of natural ventilation over the aero-dynamic metal structures.

Disadvantage

- Availability of quality, matured bamboo is a serious problem, as our bamboo production and harvesting follow the rule of forest. Till date no organized and modernized package of practice, for production and, harvesting is available. Thus procuring only 3 years bamboo of desired variety is a real problem.
- Availability of properly trained craftsmen and availability of proper bamboo curing and treatment facility is another disadvantage for this relatively new technology to implement.
- On-site improvisation/fine-tuning of the construction work is a problem for regimented work-force, It is important because for a single mistake the structure may be damaged.
- Finally and most importantly, improper and poorly-scheduled maintenance may create serious problem/disadvantage to such bamboo greenhouse.

5.10. A Success Story

For the urgent need of Agriculture department of Govt. of Sikkim this entire innovative work was done in the middle of the year of 2011 through a company known to me. The Govt. of Sikkim was doing it since few years but facing serious problem with design and strength of the structure.

After doing initial experimentation I have to build two models in Bermiok Horticultural farm of Govt of Sikkim in association with Sanhit Biosolution Pvt. Ltd. That was an instant hit and the said company got an order of 1100 nos of such bamboo greenhouse (134sqm each) to be constructed in the farmer's field scattered all over Sikkim in the year 2012-13. Cane & Bamboo Technology Center (CBTC), Jorhat, also got an order from Sikkim Govt to construct 1000 no of bamboo greenhouse of same size, with their design & technology, which they were not able to complete. On the other hand, on farmer's demand, due to superiority of our design, strength of structure, and quality of work, the company associated with me got further order of 600 nos of that innovated model.

It was a grand success to implement 1700 nos of bamboo greenhouse with the help of the above mentioned package of innovated technologies.

Just imagine the IMPACT of bamboo greenhouse on society and environment. When a total 2300 no of small 'bamboo greenhouses' were made by Govt. of Sikkim in the year 2012-13,

- It helped 2300 farming family to go for precision farming, which otherwise will not happene.
- It helped financially the numbers of bamboo growers lived in remote areas,
- It saved 2000 lakhs of INR of Govt. exchequer by not going for metal structure,
- It saved about 3450 MT of GI-pipes or 6270 MT of iron ore for our next generations.
- Finally it made a big impact on carbon footprint by using Bamboo against steel.

Apart from the above the much better and bigger gutter connected bamboo Greenhouse was implemented successfully in Nadia District of West Bengal, India in 2015.

5.11. Award and Recognition

This innovative Bamboo Greenhouse Technology, created in 2011, was awarded in India innovative Growth progremme (IIGP), 2013.

In 2011 four application was submitted for necessary patent in India on 1) Bamboo curing, 2) per-cast grout, 3) Kit for designed bamboo greenhouse, and 4) Fan-tube ventilation system. It was published in 2016 and final sanction is awaited.

After winning the above-mentioned award, this innovative bamboo greenhouse technology was invited by Global Forum of Innovative Agriculture (GFIA) to present it in its first programme at Abu-Dhabi in 2014.

IIGP (Indian Innovative Growth Programme) Award of 2013 Given for Bamboo Greenhouse Technology

This innovative bamboo Greenhouse technology was invited to be presented in the conference of Global Forum of Innovations for Agriculture (GFIA) at Abu Dhabi on February 2014 (www.innovationsinagriculture.com).

6

Climate Control Mechanism

The greenhouse technology is all about the controlling of ambient climatic factors to create a favourable 'micro-climate' for the crop growth. The magnitude of such control may differ from minimal to maximum. It may control single (like rain) to multiple aspects (like rain, temperature, humidity, sun-light, wind, carbon-di-oxide, etc in combination). It can simply be done by manipulating natural factors using the suitable/appropriate design of the greenhouse or can be done in complex form by using/installing different devices and packages (may be software) with necessary automation in a specifically designed greenhouse.

It is a vast subject and very difficult to have a thumb-rule for such climate control mechanism due to seasonal climatic variations and most importantly due to the inter-relation effect between each and every climatic factors. Thus, it should be given sufficient space to maneuver the mechanism designed for a specific greenhouse targeting a specific crop.

Ways to control these climatic factors in a greenhouse

There are two basic ways to control the climate inside greenhouse.

Primarily, through a suitable design of the greenhouse itself, the inside micro-climate can be controlled, obviously up to a certain level. Through manipulation of the design, particularly the naturally ventilated system, the rain water, temperature, humidity, CO_2 wind, etc can be controlled up to a certain level in favour of the crop.

Secondarily, by introducing equipments, tools, devices, etc we can control the micro-climate of a greenhouse. In most cases, it can take care of a number of climatic factors, like temperature and humidity, sun-light and temperature, etc simultaneously to provide a crop-wise micro-climate inside.

Frequently we use both the basic controlling methods, in combination, to achieve the required climate control system in a particular greenhouse.

We have already discussed about the different climatic factors that affect the crop micro-climate. Now we will discuss the basic control mechanism of each and every climatic factors and the related technology that can be utilized for greenhouse.

6.1. Light Control Mechanisms

The management of light is important for crop production as it has a direct relation with photosynthesis and photo-periodism, which alone or in combination optimize the yield in accordance to the crop's genetic potentiality. When other environmental factors like CO_2, temperature and water are optimal, the most favorable light intensity and duration can maximize the yield of a crop.

Plants respond in various ways to the intensity and duration of light. Thus, we should look at each of the ways of those greenhouse grow-lights that affect crop growth. Light intensity and duration must be enough for the photosynthesis needed to keep the plant growing and producing. However, all kinds of plants have not the same need.

The need for light can differ from one plant species to another and even from one cultivar to another. The successful greenhouse grower will be mindful of the various crops being grown in the greenhouse and their individual lighting needs, and will ensure crops have sufficient spacing to receive the available light. All other plant-needs can be met in the greenhouse, but unless light is carefully evaluated and managed, plants will not grow properly and yield will be lower or fail entirely.

In general, the day and night cycle actually manage the duration of light in most of the cases. In a few cases, in greenhouse, we try to extend the day-length or reduce the night length by introducing light inside when it is dark outside. On the other hand, we try to reduce the day-length or increase the length of night by covering crop to create artificial darkness. Now we have to discuss some basic feature of light, which we have to control.

Intensity of light: In general, the crop becomes light saturated in 3000 to 10,000 foot-candle (fc) depending upon the position of canopy. Three mechanisms are required to provide proper light intensity for (i) Maximization of sunlight in greenhouse, (ii) Reduction of sunlight at daytime and (iii) Provision of supplement light at poor sunlight condition. It shall be done depending upon the crop-climate combination of such greenhouses.

BOX - 12

Light saturation: It is the level of light intensity for a crop beyond which photosynthesis does not increase. Many crops become light saturated at 3000 fc assuming that all leaves of that crop are exposed to this intensity. But, it is not possible in the field as upper leaves cast shadows on lower leaves and so on, thus reducing the light intensity in the lower leaves. An individual leaf at the top of the canopy may light saturate at 3000 fc, but the whole plant/canopy may not reach light saturation until 10,000 fc. However, many of them become light saturated in 2000 fc and over 3000 fc burning starts. Some of the foliage plants are so sensitive that at 1000 fc or more light intensity they lose chlorophyll. Excess or high light intensity may cause chloroplast suppression, petal or soft tissue burning etc. in several crops, however, such intensity varies considerably from crop to crop.

Duration of light: Most crops require full sun, which is **six to eight hours** of direct sunlight daily. Plants that grow in partial sun/shade only need **three to four hours** of direct sunlight. Accumulated light within the greenhouse is often measured over the period of a day. Duration of light is expressed by day-length or, more accurately, the length of the night or dark period. All plants are not triggered by day length, or length of the dark period, to go into the reproductive growth stage. However, the plants whose growth is dependent on duration of light are divided in to two groups, viz 'long-day' and 'short day' plants (more accurately 'long-night' and 'short-night' plants). In greenhouse, growth of such crops can be monitored by controlling the day-length or night-period.

6.1.1. Maximization of sunlight in greenhouse

By increasing the penetration of natural light through covering material, to maximum possible extent in the low light periods, can maximize the growth of a crop. This can be done in the following way

1. Design of greenhouse: The thinner structural frame and widely spaced frame of the roof allow greater light intensity inside. The white paint or any reflectant (silver) colour for the frame reflects light into greenhouse instead of absorbing it. The pipe-frame Quonset and all metal gutter connected poly-houses are very good in terms of minimal frames.

On the other hand, greenhouses with curved roofs have better transmission capacity than that of greenhouse with straight roof of 25% slopes.

2. Selection of appropriate Covering material: The covering material is another vital factor to transfer the light into the greenhouse. For example glass, single layer PE-film, double layer PE-film has 90%, 88% and 77% capacity to transmit the sunlight inside.

3. Cleaning of covering material: The transmission capacity of covering material reduces with time due to deposition of dust and algal proliferation. The first shower of pre-monsoon rain cleans the dust to some extent but a permanent stain remains. That stain in combination with algae reduces the transmissivity of the greenhouse film. So cleaning of the covering material before low radiation (cloudy) period will help to maintain the transmissivity of the covering material.

BOX – 13

Effect of dust deposited in greenhouse roof

Deposition of dust reduces the entry of solar radiation inside a greenhouse covered with transparent glazing material. This reduction may be calculated by the following formula, developed by Ei- Shobokshy and Hussein (1993).

$$I_{in} a I_{out} + b I_{out}^2$$

Where I_{out} = Inside solar radiation, I_{out} = outside solar radiation,

a & b are the constant available m the following table.

Dust deposition (mg/m²) Constants	CLEAR	1.7	6	12	34	110
A	0.65	0.45	0.32	0.26	0.31	: 0.3o
B	4×10^{-4}	6.5×10^{-4}	8×10^{-4}	7.8×10^{-4}	6.7×10^{-4}	6.8×10^{-4}

---

4. Spacing of windbreak : The windbreak is placed in such a way that it should not cast shadow in the greenhouse.

6.1.2. Reduction of sunlight in Greenhouse

The reduction of light inside a greenhouse is required for two basic reasons. First is to reduce the entry of excess solar radiation inside by way of providing shade. The second is to block the light completely to induce darkness to enhance the night-length or reduce the day- length to manipulate photoperiodism in favour of short-day plants.

1. Shading Method: Shading can be accomplished in two following ways:

(i)   By spraying a shading compound on the roof (on the film) of the greenhouse.

(ii) By installing a screen fabric (shade-net) over the greenhouse-roof or inside the greenhouse-roof.

(i) Spray Method: This method is less expensive and temporary/seasonal in nature. The readymade commercial shading compound may be purchased or can be made by mixing white latex paint with water; 1:10 provides a heavy shade while 1: 20 provides a standard shade. Sprayer can be used to spray this material. Most of the shading compounds wear off by rain easily; if it does not, it needs to be washed off on the onset of monsoon season. The main problem with this type of shading is that it will provide continuous shading for a prolonged and specific period (for detail see 6.2.1.2.a).

(ii) Shade-net Method: When shade is desired for a specific stage of growth or when shade is required for a few hours (11 am to 3-4 pm) in a very hot day, it is used to cut down the excessive solar radiation to avoid temporary wilting as well as to reduce inside temperature. This method of reduction of light is very much common in tropical and sub-tropical areas and effective to avoid the negative impact of excess solar radiation. The PE-shade net can be applied in different densities of weave providing different shade values from 25 to 75%. It can be pulled off and on as per necessity.

This shade net may be applied over the greenhouse with a reasonable gap or below the greenhouse roof at gutter height. In first case this will reduce the temperature more effectively while providing shade to the roof. This system is applicable in hot and high solar radiation areas. However fixing and removing of shade-net placed over the greenhouse roof is a difficult job, particularly against strong wind. The application of shade-net below the roof is primarily for providing the shade to the crops and its effectiveness to reduce the temperature is secondary.

Instead of PE-shade-net, aluminum/steel shade-net, called as "thermal screen", may be used to provide shade. It is costlier, but the reflecting property in particular and its durability make it technically a far better shading material than PE-shade-net. Apart from shading, due to its reflecting property, it effectively reduces the temperature in daytime. It also effectively cuts off the heat loss through roofing material by reflecting back the re-radiation and maintain the higher temperature at night.

2. Long-night Method: In short night-periods, which generally occur in summer, to initiate growth of short-day plant in respect of flower and bud initiation a black cloth may be used over the crop in late afternoon. It induces darkness well before it actually happens in nature. This can have harmful effects if heat builds up underneath. Thus, in hot and humid areas the black cloth cover may be applied in late night and pulled off at late morning as per requirement. This method can effectively increase the night length by 2 to 3 hr. easily to make it 12 hr. or a little more. The black cloth may be dense enough to reduce the light intensity beneath 2 fc when the intensity outside is 5000 fc. This method should be applied everyday throughout the period until the requirement is over. This treatment is commonly used in flower growing greenhouses for certain flowers like chrysanthemum.

6.1.3. Supplement Lighting in greenhouse

It is the opposite method of long-night treatment. During relatively darker seasons, when the light intensity inside falls below the optimum level, i. e. during monsoon, and/or when the length of night is more than required, the application of light inside the greenhouse can induce the growth of crop and/or improve the quality of the crop. The intensity of supplement lighting will be about 2000 to 11000 lux (185 to 1000 fc).The period of lighting varies according to the requirement.

Several types of lights/lamps can be used in greenhouses. Incandescent (tungsten filament) lamps/bulbs are used when lighting as well as heating is required. Otherwise the excess heat emitted by this lamp makes them unacceptable. Fluorescent lamps/tube (cool-white light) is used in small greenhouses, nurseries and growing chambers. Generally commercial greenhouses do not use such kind of lamps. In Europe these lamps are used for cucumber and tomato crops to provide an additional 1000 and 2000 lux (95 and 185 fc) light respectively.

There are a number of other fluorescent lamps with special phosphorus for emitting light of higher wavelength (tube-light predominated by blue light) that helps in photosynthesis. There is a problem with this tube lighting system. Generally large number of tubes are required to obtain the desired light and, thus, it casts shadow in the daytime. Earlier High Intensity Discharge (HID) lamps like High-pressure Mercury (HPM) lamp, high-pressure sodium (HPS), and low-pressure sodium (LPS) lamps were used for supplement lighting in greenhouses in different parts of the world. Out of these lamps LPS lamps are most efficient for supplementing greenhouse lighting. These lamps are available in 35, 55,135 & 180 watt sizes and 27 percent of electrical input into lamp and ballast is converted to visible radiation.

In recent time, the energy efficient compact fluorescent light (CFL) and light emitting diode (LED) are the two most favorable options for greenhouse lighting.

The most common CFL grow lights are 25 watt bulbs that can be used in standard household bulb socket or larger 125 watt mogul base bulbs. The 25 watt CFLs are pretty small and most often used for growing single plants or for seedling until it grows a little larger. It normally requires four 25 watt CFLs to grow a healthy medium size plant. CFL Grow Lights typically have two spectrums – 6500k for growing and 2900k for blooming. Small CFLs like a 25 watt are usually available in 6500k only. The 125 watt models are normally offered in both 6500k and 2900k.

LEDs have a unique efficacy advantage in horticulture. Plants appear green because they absorb red and blue, the band-gap energy of the two primary photosynthetic reactions. With LED lighting, the color of the light can be tuned to "horticultural red" (660 nanometers). The LED lamps are available in the market in a customized model for greenhouse lighting.

On energy front an LED luminaire, for example, could put out red and blue photosynthetically available radiation (PAR) slightly greater than a standard 1,000-watt HPS luminaire, while consuming only 325 watts. The PAR unit of measurement is standard in horticultural lighting, since it is weighted for plant photosynthetic response. Thus, the PAR is expressed as the amount of energy by which the instantaneous light incident upon a surface and is measured by micromole per square meter per second ($\mu mol/m^2/s$). Here 1 mol is 6.02×10^{23} photons and 1 micromol (μmol) is 1-millionth of a mol.

Example: How much light is supplied by a high pressure sodium lamp that generates 100 $\mu mol/m^2/s$, if kept on for 12 hours?

- Multiply instantaneous light by 3,600 seconds/hour and multiply by the number of hours on.
- Then divide by 1 million (1,000,000)
    - 100 $\mu mol/m2/second$ x 3,600 seconds/hour x 12 hours = 4,320,000 $\mu mol/m2/day$.
    - 4,320,000 $\mu mol/m2/day$ divided 1,000,000 = 4.32 mol/m2/day

Commercial application: In tropics and semi-tropics, the area of high intensity solar radiation, shading is frequently used commercially to optimize the light for growth of the crop. Shading may be applied on greenhouse cultivation temporarily, for a few weeks or a few days or even for a few hours, depending upon the crop-climate requirement. To avoid 'temporary wilting' due to high rate of transpiration caused by high solar radiation the shading in greenhouse is mostly used.

Supplement lighting is not a common practice in tropical and sub-tropical greenhouses. However, this method is used frequently in countries like Canada, Northern Europe, which lie at 50°N latitude and above where light intensities are not favorable for plant growth. However, in case of growing of seedling this mechanism may be useful in cloudy or low radiation season and in long night period.

The supplement lighting technology also opened a new option for indoor gardening/cultivation, which recently gained popularity in urban areas. This indoor farming system brings a new dimension in 'urban farming', which is otherwise associated with small or mini greenhouses.

6.2. Temperature Control Mechanisms

Over last 25 years different greenhouse technologies, intending to maintain a favourable inside climate, has come up. Most successful and economically viable know-how to control greenhouse temperature are (i) passive-heating and (ii) passive-cooling technology.

The passive heating technology is the method of using the storing/conserving capacity of the solar heat by the greenhouse utilizing the 'greenhouse effect'. On the other hand passive-cooling technology is the method to utilize wind and shade to reduce heating effect of solar radiation. Hence, two basic types of temperature control mechanisms are used in greenhouse technology. First is the temperature reducing i.e. cooling mechanism and the second is the temperature increasing i.e. heating mechanism.

Up to 18th century temperature reducing or cooling problems were considered as secondary in respect to the greenhouse heating problems, as because the greenhouse technology, till then, was restricted in cooler climates. With the spread of greenhouse cultivation towards equator, i.e. in sub-tropical and tropical areas, the problem associated with cooling became relatively more important and now it is a major cause of concern in such areas.

During the past decade, study of naturally ventilated greenhouse received maximum importance in respect of greenhouse-cooling research.

In warmer climates, generally termed as tropical and subtropical zone, which are placed between 40°N to 40°S latitude and receives high intensity solar radiation, the temperature reducing or cooling mechanisms are used predominantly for greenhouse cultivation.

On the other hand, places above and below 40°N and 40°S respectively, where cool climate prevails and which is termed as temperate and cool climate, require increase of temperature i.e. heating mechanism for greenhouse.

However, the above division is not always applicable in all the cases. Different microclimatic factors, like higher altitude and coastal climate, shall be considered while selecting the temperature control mechanism. In some places (between 23^0 to 40^0 of both North & South latitude) extreme levels of temperature prevails in day and night or in winter and summer. That situation compels to adopt both the temperature reducing as well as increasing mechanisms for a greenhouse. Now we will discuss the details of these mechanisms.

6.2.1. Temperature reducing or cooling mechanism

As most of the cultivable areas of the world are located between 40°N and 40°S latitude, the temperature reducing mechanisms are gradually getting more importance in greenhouse technology. Three basic types of mechanisms are used for greenhouse cooling either as a single measure or in combination depending upon the requirement of crop-climate factors. These mechanisms are (i) ventilation, (ii) shading and (iii) evaporative cooling mechanism.

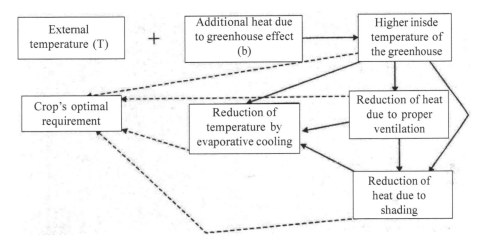

Fig. 37: Different ways of heat-flow chart to manage greenhouse temperature.

We all know that the inside temperature of a greenhouse is always 10-20% higher than the ambient temperature due to 'greenhouse-effect'. We can also assume that proper ventilation and air circulation can effectively reduce the additional temperature generated inside a greenhouse. So the inside and outside temperature of the greenhouse becomes almost same. Now if this temperature is still 10 to 20% higher than that of the optimal temperature required for the crop, then it can be further reduced effectively by way of shading or thermal screening. This level of cooling is possible only with moderately high outside temperature. But if the outside temperature is too high even after providing proper ventilation and shading of the crops, further reduction of temperature is required. This can only be done through any of the evaporative cooling methods described afterwards (6.2.1.3.).

6.2.1.1. Ventilation

Ventilation is a necessity for any type of greenhouse. The size, type, position of ventilator or window may differ from greenhouse to greenhouse. These ventilators or windows are permanently open or may open temporarily i.e. have closable arrangement. Ventilation serves two main functions for greenhouse cultivation: (a) Exchange air (or gas/CO_2 and water vapour) between interior and exterior of a greenhouse and (b) cooling of greenhouse by ventilating the hot air built up due to the greenhouse-effect. At present we are interested on the cooling aspect only.

Without ventilation a greenhouse is just a solar furnace, particularly in high solar radiation. In general a closed greenhouse with transparent covering will produce 15°F to 20°F or more additional heat inside while compared to outside. Thus cooling a greenhouse on a hot day is a very difficult task, which cannot be accomplished without a good ventilation system.

A few types of ventilation system are practiced in greenhouse technology. However, some of these types are used in combination to increase ventilation efficiency.

(i) Natural Ventilation

When the exchange of inside and outside air takes place with natural air current/ flow through ventilator/window of a greenhouse, then it may be called as natural ventilation. In greenhouses a potential gradient of air pressure between inside and outside develops due to temperature difference. This potential gradient helps in air exchange in association with natural air flow or wind.

However, wind speed has a significant effect in naturally ventilated system. Non-uniformity of wind speed is another factor that creates problem in naturally ventilated greenhouses because wind speed varies widely from 0.35 to 6.7 mile/ second (Seginer et al., 2000). Wind direction may also affect ventilation rates in naturally ventilated greenhouses.

Canopy size is another factor that affects the ventilation efficiency particularly in naturally ventilated greenhouses. It was found that a crop of tomatoes with 1.5 to 1.7 m height could decrease ventilation efficiency in naturally ventilated houses by as much as 28% (Boulard et al., 1997).

The position and size of ventilator/window of a greenhouse plays a big role in managing these two problems (non-uniform wind and higher canopy) related to inside airflow or natural ventilation. Furthermore, in case of putting net to the windows the size of the windows has to be re-calculated.

Ventilator at top/ridge of the greenhouse: These types of ventilators are placed at the top of the roof structure along the roof. Depending upon the type of roof structure it may be placed in a suitable position (Fig. 14.a & b) that allows the hot ascending air to pass easily out from the greenhouse. Sometimes these ventilators have closable system to conserve heat in winter seasons or in cold nights.

In warmer climate, the side facing ridge or roof ventilators (saw-tooth and ridge & furrow greenhouses) may be placed in windward direction, thus the wind current creates a negative air pressure at the top of the opening and sucks the warm air from the greenhouse and reduces the air pressure inside, which drives in the ambient air through windows placed in side-wall.

In taller greenhouses, the ridge/top ventilators will increase its efficiency by way of creating the 'chimney effect'. In such greenhouses on an average the roof vents are of about 10 percent in size of the floor area of a greenhouse.

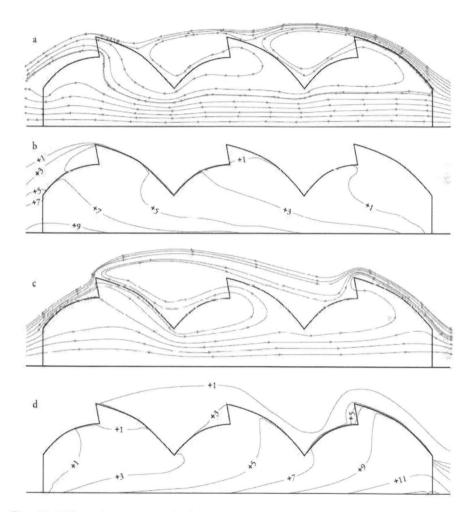

Fig. 38: Effect of natural ventilation on temperature – a) Path lines of eastward wind, b) Corresponding isotherm, c) Path line of westward wind, d) Corresponding isotherm

Ventilators at sidewall and gable of the greenhouse: In hot climate, only roof/ ridge vent is not sufficient to create necessary ventilation to control the rising temperature inside the greenhouse. In that case the natural airflow or wind should have direct entry to the greenhouse to facilitate equilibrium between inside and outside temperature. This can be made possible by providing sidewall windows designed in proper places.

Size of the sidewall ventilator: The total size of sidewall ventilator or opening may vary from 10 percent to as high as 80 percent or more of the floor area. However, the said percentage is dependent on size of greenhouse and the corresponding peripheral length. Both sidewalls along the width of a rectangular greenhouse may be kept open to have a better air movement and to create the 'tunnel effect' inside. Depending upon the crop-climate requirement, the position of sidewall openings may be changed. It may be placed in lower or middle part of the side-wall depending on the crop. To obtain best possible ventilation efficiency such windows may have the facility to close or open whenever necessary.

In tropical wet climate where day-night or winter-summer temperature differences are not significant, the entire sidewall may be kept open permanently to achieve very high level of ventilation inside the greenhouse. These wide sidewall openings, used for uninterrupted airflow, may have the closing facility with insect proof net to avoid pest and subsequent disease problems.

Assessment of air exchange rate and ventilation rate of a greenhouse:

To assess the ventilation area of a greenhouse, the estimation of air exchange rate and the ventilation rate are to be determined first. This may be expressed in the following way:

$$\text{Air exchange value of a greenhouse} = \frac{\text{Exchanged air volume per unit time}}{\text{Volume of the greenhouse}}$$

It is generally expressed as liter/hour (l/h) or cubic feet/minute (cfm) and the approximate value of this at which sufficient ventilation of a greenhouse is expected may be 55 l/h or 0.0324 cfm. However, this requires to be increased with increase in solar radiation. This standard figure consider inside light of 5000 foot-candle (fc).

$$\frac{\text{Air exchange rate per unit floor area}}{\text{of a greenhouse (ventilation rate/ARR)}} = \frac{\text{Exchanged air volume per unit time}}{\text{Area of the greenhouse}}$$

It is generally expressed as $m^3/h/m^2$ or cfm/sft. The approximate value at which sufficient ventilation of a greenhouse is expected is 170 $m^3/h/m^2$ or 9 cfm/sq.ft.

However, this air exchange rate depends on altitude, inside light intensity and rise of inside temperature or temperature difference between fan and pad of fan-pad cooling system of the greenhouse. This ventilation rate may also be termed as air removal rate (ARR).

The rate of air removal from the greenhouse must increase as the altitude of the greenhouse site increases in order to have an equivalent cooling effect. The rate of increase of air exchange rate may be determined by multiplying with 0.04 for every increase of 304.8 m or 1000 ft elevation up to 5000 ft and then the factor will increase @ 0.05 for every increase of 1000 ft. (Table 16).

Table 16 : Air removal rate of greenhouse in different elevations.

Elevation in feet	Below1000	1000	2000	3000	4000	5000	6000	7000	8000
Rate of air removal in cfm/min	9.00	9.36	9.72	10.08	10.44	10.80	11.25	11.70	12.24

Developed by NGMA, New York, USA.

In respect of light intensity, the standard air exchange rate i.e. 170 m^3/h/m^2 or 9 cfm/ sft is considered at 5000 foot-candle (fc) inside the greenhouse. Now for every increase or decrease of 500 fc of light the rate may increase or decrease respectively to the extent of 0.1 factors less or more and thus doubled at 10000 fc of light intensity (Table 17).

Table 17: Air removal rate of greenhouse in at different light intensity at day-time.

Light intensity (fc)	4000	4500	5000	5500	6000	6500	7000	7500	8000	9000	10000
Rate of air removal in cfm/min.	7.2	8.1	9.0	9.9	10.8	11.7	12.6	13.5	14.4	16.2	18.0

Developed by NGMA, New York, USA.

The Area of Ventilation (Ventilation ratio): It is generally expressed as percentage of ventilation area against the floor area of a greenhouse. In any case the total ventilation opening of a greenhouse should not be less than 20 percent of the floor area. In greenhouses of cooler climate with closable ventilators the figure may be around 20 percent but in case of warmer climate it may be increased to 60 percent and sometimes up to 100 percent or more. The experiment of V.M. Salokhe *et al.* (2006) shows that out of 20, 40, 60, 80 and 100 percent ventilation area against the floor area, 60% is the minimum requirement in order to maintain favourable microclimate for the crop in tropical greenhouses.

Increase in internal temperature has less effect on greenhouse with full crop. With 20% ventilation, the increment of inside temperature is about 1.5°C for full crop and 4.5°C at empty condition. The humidity increases by 37.5% and air exchange rate decreases 40% with 20% ventilation in comparison to 100% ventilation.

If the ventilation is designed at the two sidewalls along the length of a single or stand-alone greenhouse structure and if the side-wall ventilation is proposed to be 20%, then the maximum width of the ventilation opening may be calculated as follows:

Width of greenhouse $= {}^{2\text{ X width of ventilator}} /_{0.2}$

The length of greenhouse is not required to be considered and there will be no ventilators in the two sidewalls along the width. If it is considered that the altitude and light intensity are within the normal range (i.e. within 300m and 5000 fc respectively), then a greenhouse of 12M width with 20 percent ventilator area will require a ventilation width of 1.2 M.

In case of multi-span greenhouse both the gables are provided with ventilation opening of the same width as in the sidewalls, to achieve 20 percent. In that case, the calculation will be as follows.

Width of greenhouse $= {}^{2\text{ X width of ventilator}} /_{0.2 - 2\text{ X (width of ventilator / length of greenhouse)}}$

Accordingly one can calculate the necessary width of the side-wall windows/ ventilators of a greenhouse with higher percentage of ventilation by using the same basic formula. In that case use 0.4 or 0.6 for 40% and 60% respectively instead of 0.2 in the formula.

Effect of Insect-proof (IP) net on ventilation opening: In most of the recent greenhouses the ventilation openings are covered with insect protection (IP) nets to protect the crops from insects particularly the disease carrying vectors like white fly etc. In that case ventilation efficiency is reduced to a fair extent. As far as the effect of screens on ventilation rate is concerned, it was found that their (screen with 0.5 $m^2 m^{-2}$ porosity) use in the vent openings caused a greenhouse ventilation rate reduction of about 33% [C. Kittas, N. Katsoulas and T. Bartzanas et al. (2005)]

To compensate this reduction of air movement the ventilation openings have to be correspondingly larger. The insect-net or insect-proof nets are of different types depending on the size of pores. There are 30, 40, 50, 60, and 70 mesh IP-net. Depending on the prevailing pest of the selected crop the IP-net is selected for a greenhouse.

(ii) Forced or powerised ventilation (exhaust fan) system

The above discussion about ventilation may raise a question that whether the natural ventilation can successfully reduce the excess heat generated inside the greenhouse in daytime? With high percentage of relative humidity and less ventilator openings, in many cases, it is not possible to reduce the excess heat inside, particularly in warm climate. Natural ventilation in combination with shade screening and application of induced ventilation, through a modified exhaust system, can successfully reduce the excess heat in a greenhouse.

(a) Normal Exhaust system for closed greenhouse and for fan-pad cooling:

In the above mentioned situation an exhaust fan may be used to expel hot air out of the greenhouse forcibly, thus creating more pressure gradient to let in the fresh and relatively cool air inside. However, the placement of exhaust fans and ventilators shall be designed properly. Depending upon the size of fan and gable, one or two fans per gable are generally placed to drive out the hot air and they are generally placed in the upper side of the greenhouse. In a very rough estimate, for $60m^2$ floor area, two-exhaust fans of 24-inch diameter may be used. To obtain proper efficiency for small greenhouse, it should take 1 minute to make a complete change of air and for large commercial greenhouse it should take 2 to 3 minutes.

Crude design of a forced ventilation system (exhaust fan): Use of an exhaust fan is to move air out from one end of the greenhouse while outside air enters the other end through inlet louvers. The size of the exhaust fans should have the capacity to exchange the total volume of air in the greenhouse within 2 minutes. The total volume of air in a medium to large greenhouse can be estimated by multiplying the floor area by 8.0 (the average height of a greenhouse). A small greenhouse (less than 5,000 ft^3 in air volume) should have an exhaust-fan capacity estimated by multiplying the floor area by 12.

1. The first thing is to determine the amount of air required. This is stated in cubic ft. of air removal per minute (CFM).
2. The calculation of CFM is based on several factors:
   a. Light Intensity - type of shade-net to be used is determined by geographic location and weather.
   b. Length of Greenhouse - from air inlet to air outlet point.
   c. Temperature Difference - from air inlet to air outlet point.
3. Use the following basic formula to determine the CFM required for a greenhouse:

CFM = Length x Width x 12

4. Increase or decrease in the CFM amount is calculated by using the following factors:
   a. Shade cloth factor - (i) 40% = 0.6, (ii) 50% = 0.5, (iii) 60% = 0.4.
   b. Length of greenhouse factor- (i) up to 25ft=0.6, (ii) 26-50ft=0.5, (iii) 51-100ft = 0.4

EXAMPLE

12' x 36' Greenhouse-with 50% Shade Cloth.

To Determine basic CFM required using standard formula. - 12x36x12 = 5184 CFM.

We must now adjust this based on the Length Factor and Shade Factor.

5184 x .5 (Shade Factor) x 1.5 (Length Factor) = 3888 CFM

This is the CFM required to properly cool this greenhouse.

If using evaporative coolers (swamp coolers), make certain that you size these units in accordance with their Actual CFM instead of the Industry CFM Figures.

Remember: The greenhouse must provide an adequate exit for the total CFM. One can use an open door, roof vents or exhaust shutters properly sized for the total CFM. The most efficient system utilizes exhaust shutters situated on the wall opposite to the air entry openings.

(b) Fan-tube winter cooling system for closed greenhouse

It is basically a forced cooling system for a greenhouse where outside temperature (at daytime in winter) is much lower than that of inside, particularly in winter day-time.

In case of winter cooling system a distribution tube (mostly PE-film tube) is placed inside the greenhouse at a suitable height along the length (Fig. 39) on which holes are present to discharge relatively cooler air uniformly (for detail see unit heating system, Chapter 5.2.2.). One side of the tube is closed and the other side acts as an inlet with louver and a blower fan installed in between the louver and the inlet point. In case of large greenhouse (more than 200 ft long)

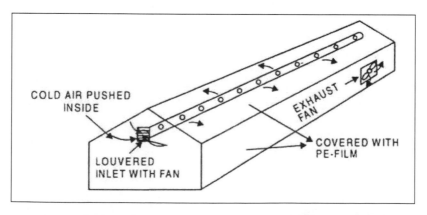

Fig. 39: Day-time Fan-tube winter cooling system for closed greenhouse.

the tube is open and sucks cold outside air through both the ends by two fans. The fans suck outside cold air and distributes it within the greenhouse through the tube evenly thus cooling down the greenhouse.

The number and diameter of the tube depends on the width and length of the greenhouse (Table -14). However, the specification of holes is designed to maintain the required airflow rate inside. Work in England by G. A. Carpenter specifies that the total area of all the holes in a tube shall be 1.5.to 2 times of the cross section area of the tube. The holes along the tube exist in pairs on the opposite vertical sides and a distance of 2 to 4 feet is common.

Table 18: Number (N) and Diameter (D), in inches, of air-distribution tubes against the size of greenhouse.

Length (ft)	50		100		150		200		250	
Width (ft)	N	D	N	D	N	D	N	D	N	D
15	1	18	1	18	1	24	1	30	1	30
20	1	18	1	24	1	30	1	30	2	24
25	1	18	1	24	1	30	2	24	2	30
30	2	18	2	18	2	24	2	30	2	30
35	2	18	2	24	2	24	2	30	3	30
40	2	18	2	24	2	30	2	30	3	30
50	2	18	2	24	2	30	3	30	3	30

Note: These recommendations are based approximately on an air-flow rate of 1700 cfm / sft of cross-sectional area of the tube.

To maintain a negative pressure inside the greenhouse exhaust fans may be used but the air exchange rate of the exhaust fans must be less or at least equal of the fanned air to maintain a continuous flow. Instead of minimum summer air rate of 8 cfm/m^2 of floor area here the rate will be around 2-cfm/m^2 floor area.

(c) Fan-tube ventilation for naturally ventilated greenhouse

It is a modified exhaust system designed for greenhouses of humid & hot climate. It was innovated by the author in 2011 and named as Fan-tube ventilation system.

As the normal exhaust fan will not be able to do the work properly due to the presence of large openings/windows around the greenhouse, this modification is essential for a proper exhaust system to act homogeneously.

It is basically an opposite system of fan-tube winter cooling system discussed earlier. However, this system can be effectively used, with some structural modification, as negative pressure system for cooling in warmer climate where day-time inside temperature is high. In such cases, if the natural ventilation system (60 % or more ventilation) is practiced and all the ventilators are covered with insect proof (IP) net, then this modification is the only way to cool down the greenhouse through ventilation. It will be achieved by expelling the hot air outside, thus creating a pressure deficit zone inside. This in turn will let in more fresh and relatively cool air through the side-wall ventilators covered with IP-net, which otherwise is not possible by ambient air current.

For this summer cooling mechanism (for greenhouse with higher percentage of ventilator openings covered with IP-net) some constructional modification in respect of the distribution tube (rigid one) may be done which will act as a suction tube. It will suck warmer air from inside homogeneously and throw it out by means of an exhaust fan. This system will increase the efficiency of natural air exchange rate and increase the ventilation rate or air removal rate (ARR) of the greenhouse. It can also maintain an effective air-circulation inside, where air takes entry from the side walls covered by IP-nets and circulate around the plants and move towards the suction tube to be blown outside. (Fig. 40).

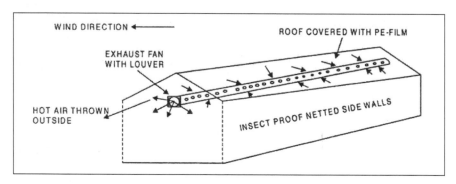

Fig. 40. Fan-tube ventilation system increase air flow rate in greenhouse protected with IP net, thus reducing temperature and humidity.

The size and power of the fan will be calculated on the basis of standard air removal rate (ARR) per floor area of the greenhouse. Thus the size and power of the fan/fans is dependent on the size (volume or floor area) and the location of the greenhouse.

The standard air removal rate (for detail see fan-pad cooling system) of a greenhouse is required to be corrected on the basis of altitude, inside light intensity, and temperature difference between inside and outside. Now the size and capacity of fan, necessary for forced ventilation in a greenhouse, shall be calculated properly. For that matter the air removal rate (ARR) is calculated considering different factors affecting the same. Table-16 provides the correcting factors for altitude (factor-A) and light intensity (factor-L). Other factor, i.e. the temperature difference maintained (factor-t), is provided in Table 19 below.

Table 19: Correction factors of ARR for maintaining a higher inside temperature over outside temperature.

Temp. diff. (°F)	18	17	16	15	14	13	12	11	10	9
Factor-1	0.83	0.88	0.94	1.0	1.07	1.15	1.25	1.37	1.5	1.67

This means, at the ideal situation, i.e. below 1000 feet altitude, with 5000 foot candle light intensity inside and a 15°F temperature difference, the said correction is not required. However, to maintain the same status with 10°F temperature difference, the ARR is required to be increased 1.5 times. Hence, the capacity of the fan will also be increased. (See example of fan-pad cooling system)

6.2.1.2. Shading

Apart from protecting the crop from high solar radiation, as discussed in 6.1.2 (II) of this chapter, shading has a great role to control the inside temperature of a greenhouse.

This method (shading) is very much specific for the warm i.e. tropical and sub-tropical areas. In these areas the average daytime (particularly mid-day) solar radiation and subsequent temperature are much higher in comparison to the optimal requirement of the crops cultivated in a greenhouse. 40-50 percent reduction of solar radiation in highest light intensity periods can reduce temperature below the shade to the extent of 20 percent of outside.

There are two methods of applying shade in a greenhouse, which are as follows:

(a) Application of shading compounds: By spraying shading compound on the outer side of the covering material of the greenhouse. This is done at the onset of warm season. The shading compounds are common slaked-lime [Ca(OH)$_2$], used for whitewash, or calcium carbonate [CaCO$_3$] or white cold-water paint or white latex paint mixed with water. In case of lime, gum or sticker is used to stick the coating properly on the plastic.

White coating of lime reduces the inside temperature by 3-4°C by way of reducing light intensity to the extent of 20-25 K lux. The average requirement of lime for moderate shading is about 40gms/m^2. All these water-soluble materials can be sprayed from outside by means of a sprayer. The concentration of the spray material will determine the type (heavy/ standard/light) of shading provided inside the greenhouse.

These coatings reflect much of the heat received from the sun and permit a lesser amount of radiation (shading) into the greenhouse reducing the inside temperature. After the season is over the shading material is washed off by first rainwater or manually. The problem of this type of shading is that it is somewhat permanent for a period of time. If for any reason the normal sunlight is reduced, the solar radiation received by the crop inside the shaded greenhouse will be reduced further thus reducing the photosynthesis rate.

(b) Application of shade-net or Thermal screen: The second method is application of shade-net or thermal screen. In general polyethylene shade-net is used for this purpose and it is applied and worked in similar fashion and is used for reduction of impact of strong solar radiation (see chapter 5.1.2.). Here the shade-net works as a screen to trap heat from a portion of sunlight before it reaches to the crop etc.

Apart from the PE shade-net, aluminum/steel shade-net (**thermal screen**) is also used for this purpose. Apart from shading it can reflect back a portion of sunlight instead of absorbing it like PE shade-net, thus enhance cooling effect inside. Baily (1998) studied the shading effect and reported that an aluminum plated mesh reduced the inside temperature by 6°C in comparison to a greenhouse without shading at an ambient temperature of 33°C.

This method of shading is of two kinds. First is the (1) application of shade-net over the roof of greenhouse and the second is (2) application of shade-net below the roof of greenhouse (Fig. 31).

Shade-net above the roof of greenhouse: Here the shade-net of desired type and specification is applied over and above the roof of greenhouse either with a gap between shade-net and PE-film or along the PE-film. The shade-net may be removable. Till now it is not a common practice probably due to the problems associated with structural designing and operational aspects in case of removal type. However, in hot humid areas this is a much better method to reduce the detrimental effects of high intensity sunlight and high temperature, received by the naturally ventilated greenhouses. In hot dry climate this type of shading coupled with any evaporation cooling system can reduce the temperature very effectively and can save energy (Fig.41.a & b).

In this method the high intensity solar radiation is cut off before it reaches the plastic of the roof. The reduced solar energy then passes through the air between shade-net and roof-covering and loses its heat energy further. Then it enters into the greenhouse with less amount of energy. With proper ventilation it can reduce the inside temperature than that of outside.

Shade-net below the roof of greenhouse: Application of shade-net below the roof, at the gutter height, is the common practice in the greenhouses available here (Fig. 41.c). It is used mainly to cut down solar radiation before it reaches the crop. It basically protects crop from excessive rate of transpiration. This will avoid temporary wilting as well as save the crop from exposure to excess solar radiation. Reduction of temperature is additional benefit and its capacity to reduce the inside temperature is somewhat less compared to the above-roof method.

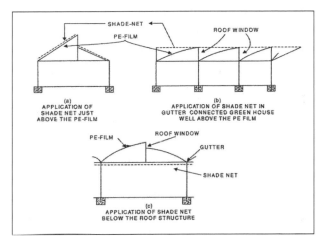

Fig. 41. Different ways of application of shade-net in greenhouse

Here the roof of the greenhouse receives and passes most of the high intensity solar radiation inside. It heats the covering material and subsequently air layer beneath the roof. As the shade-net is a highly porous material, the air layer below it has ample scope to come in contact with the hot air layer above the net. Gradually the air below the shade-net also gets heated, so the temperature of the entire greenhouse will be in the higher side if compared to the above-roof method.

However, in case of below-roof method if an exhaust fan is placed between the roof and shade-net and the hot air layer in between is continually driven out of the greenhouse then this method may reduce the inside temperature effectively.

6.2.1.3. Evaporative Cooling

Evaporative cooling systems are the methods practiced to cool down the temperature of the greenhouse by conversion of sensible heat into latent heat. Since evaporation occurs only at water air interface, it is best to provide as much surface area as possible between water and the air. However, low humidity is a basic pre-condition of effective evaporative cooling system. With more than 80% humidity the effect of such cooling system turns to be less effective. In that case air-circulating fan will be effective.

Three basic methods of evaporative cooling are used in greenhouse technology: (1) Air circulation around plant surface, (2) Application of water through sprinkling/misting/fogging and (3) Fan-pad cooling system. Out of these three methods the 1st one is indirect method and the next two are direct methods. Any evaporative cooling method is a very simple process to reduce temperature if the humidity level is low. While water molecule evaporates, it extracts the heat from its surroundings, thus reducing the temperature. This heat is termed as latent heat. However, very high humidity reduces its efficiency, actually makes it ineffective, while low humidity increases the same significantly.

(i) Air Circulation by 'air-circulating fans'

Installing circulating fans in a greenhouse is a good investment. The 'air-circulating fans' increase the flow of inside air directly, particularly around the leaves, which increases the evaporation as well as transpiration and cool down the temperature of crop. The 'exhaust fan' or exhaust system, to some extent, do the same thing indirectly but is not much effective in high humid situation.

In humid and hot climate, when ambient temperature and humidity are in higher side, it is very natural that both these factors have a tendency to increase further inside a greenhouse. Under such conditions 'air-circulating fans' installed inside a greenhouse will do a good job to reduce the harmful effect of high humidity and temperature on plant. The increased airflow around the plant canopy reduces the leaf temperature and disperses the high humidity around leaves, which maintain the transpiration pull of the crop. This will work best when coupled with exhaust fans that will throw out the accumulated hot and humid air and allow the fresh air to let in from the greenhouse window.

In cool climate, during winter in day-time, when the greenhouse is heated, you need to maintain air circulation in such a way that temperature remains uniform throughout the greenhouse. Without air-mixing fans, the warm air rises to the top and cool air settles around the plants on the floor.

During rainy season, when humidity is high and high ambient temperature cools down due to rain, this air circulating fans may be used judicially to disperse the higher humidity around plant canopy.

Design of air-circulation: Small fans with a cubic-foot-per-minute (ft³/min) air-moving capacity equal to one quarter of the air volume of the greenhouse are sufficient. For small greenhouses (less than 60 feet long), the fans are placed in diagonally opposite corners but away from the ends and sides. The goal is to develop a circular (oval) pattern of air movement. The fans are to be operated continuously during the required period of a day.

(ii) Application of water directly into the greenhouse

It is the simplest method of reduction of temperature using latent heat absorbed at the time of evaporation of water. However, selection and design of the method is a difficult job as because its efficacy is dependent on local climate, inside micro-climate, and pathogen-crop relation. The brief description of these methods are given below.

(a) Sprinkling of water: Direct application of water manually or by machine inside the greenhouse may reduce the temperature to some extent. The efficiency of this method is considered low as high amount of water is required for this method. The main other constraints of this simple method is that, in general, the crops cannot withstand/grow properly in the wet or soggy condition evolved due to this process. This wet condition is also favourable for disease infestation. The larger size of water drops is a major problem of this system. Bigger water drops produce less water evaporative surface while in the air and after touching the crop or ground a bulk amount is percolated down to the soil. In high humid condition its efficacy is reduced further.

Using this principle, some other method of water application in greenhouse has come up which have a very high evaporative cooling effect. These are misting and fogging which is discussed separately.

(b) Application of water on the roof of the greenhouse: Application of water on the top of any structure to reduce the temperature is a known fact. This can be done with the help of mini sprinkler. In hot areas, where water is available, water may be applied on the roof of a greenhouse to reduce the increase in inside temperature. But, poor water holding capacity and slope of the roof makes the idea less effective. Where shading is required at mid-day periods in hot summer and further cooling is necessary to reduce the temperature further, this principle of roof evaporative cooling may be adopted with some modification.

In the above situation a perforated pipe may be placed on the top of the ridge and the roof may be covered with removable jute cloth knitted with sufficient gap in between threads and having light transmission capacity. The perforated pipe emits water at a low rate, just more than the evaporating rate. The heat is dissipated from the roof of the greenhouse air and cause water to evaporate from the gunny net, which in turn, provide cooling effect. Continuous evaporation of water from roof lowers the temperature of the roof, thus reducing the inside temperature significantly.

(c) Misting: It is the mechanism to disperse water in the form of fine droplets ideally measuring 50 to 100 microns (0.002 to 0.004 inch) in diameter. In misting, the droplets are air borne for a very limited period and falls on the ground or crop due to gravitational pull. In general less than 40 micron droplets remain suspended in the air and more than 100 micron is categorized as rain.

The misting system has been introduced in greenhouse since 1940s, mostly for nurseries, particularly for 'mist-chamber', where this was used to irrigate the plant materials and maintain the high humidity. Now with expansion of greenhouse technology in the dry warmer climates misting is used in greenhouse cultivation as evaporative cooling method, which also maintains the desired level of humidity inside (for detail see chapter 6.3.).

Spraying of fine water droplets on the crop canopy, through misting, creates a film of water on the leaf surface. This intercepts the irradiation and helps to evaporate the water from the leaf surface thus reducing the leaf temperature. This also reduces the vapour pressure of the leaf by way of cooling down the leaf and lowers the leaf to air vapour pressure gradient thus slowing down the transpiration.

The misting system consists of four different segments e.g. (1) mister, (2) main and sub-main water lines (3) pump, filters etc. and (4) system controllers.

In greenhouse misting system, pressurized water is supplied through pipes to an emitting device called mister or mist-nozzle. Misters are commonly of two types: (1) deflection type and (2) whirl type.

The deflection-type of mister consists of a small opening or 'orifice' wherefrom the pressurized water comes out with high speed and a firm flat object or 'anvil' is placed just above the orifice. The high-speed water coming out from the orifice strikes the anvil, breaks and deflects as finer droplets and disperses in the form of mist. In general these misters are made of quality metal. Now-a-days excellent hard-plastic misters/nozzles of similar kind are available in the market. This type of misting system is cheaper and easy to maintain but uses more water in comparison. This produces water droplets of different sizes thus the homogeneity of the spray is less.

However, the adjustable 'anvil' makes these misters multi-functional and provides scope to monitor size of droplets (coarse or fine) to match with different purposes like irrigation, humidification and cooling.

In Whirl-type mister, the water under high pressure is forced through curved internal grooves of specially designed mister/nozzle which produces mist while coming out from the orifice of the nozzle. Many of these whirl-type misters have water output of 9 to 20 liters/hour. This produces a more homogeneous spray of specific range of fine droplets.

Controlling of misting system- Capacity and efficiency of these systems can be controlled or monitored by manipulating the necessary aspects of the system, which are 1) type of mister, 2) spacing of misters, 3) placement of misters, 4) misting period, and 5) frequency or scheduling of misting.

Type: The type of mister and operating pressure will determine the quantity of water to be sprayed in unit time. The size of average droplets will determine the efficiency of the misting system; finer the size of droplets higher the efficiency of the system.

Placement: In general the misters are placed above the crop in such a manner that the system can uniformly cover the cropped area of the greenhouse.

Spacing: The specing of mister will depend on the distance of horizontal spraying of the mist, the height of its placement and the height of the crop.

Misting Period: An intermittent misting for a very short duration, ideally 5 to 15 seconds, is generally planned. For larger greenhouse or nursery mechanized **travelling boom system** containing mister/nozzle are used.

Frequency or scheduling of misting: For proper control over misting in greenhouse automatic control devices are necessary. Electrically operated timer mechanisms are available that operate the misting at desired intervals. The clocks for regulating the application of water, e.g. 5 seconds ON and two minutes OFF, may be used. It is easy to install, inexpensive and dependable. However many other modern devices are also available to control the misting system in variable manner through necessary software programme.

(d) Fogging: Fog in true sense is the concentration or a collection of ultra fine water droplets, of 2 to 40 micron size, suspended in air. So fogging is the system by which fog is generated artificially to modify a micro environment. Fogging is a specialized system for greenhouse, as it will not work properly in open field due to wind.

Like misting system, fogging system also passes water through the pipes with high pressure, generated by pump, which is connected to fog-generator or atomizer or 'fogger' that produces fog in the greenhouse. However, common types of foggers produce both fog and micro-mist.

The fogging system scatters very fine water droplets in the air and because of its finer size the drops remain suspended in air until evaporated. At the time of evaporation the water draws heat (latent heat) from the air reducing the inside air temperature significantly and instantly. This system increases both the relative and absolute humidity gradually.

The fog in greenhouse maximizes vapour pressure of the air by raising the ambient humidity and lowers the air to leaf vapour pressure gradient reducing transpiration, which is not good for the crop. Hence it is not a viable option for high humid climate.

Proper fogging system, particularly the system with high-pressure fogger, does not wet the crop/plant or soil of greenhouse. The key to successful fogging hinges on a good ventilation system to avoid heat build up from stagnant air.

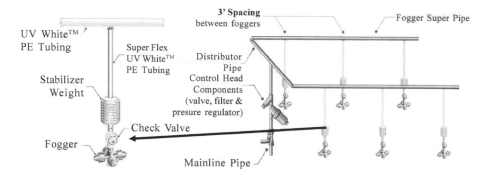

Fig. 42. Fogger and the fogging system for greenhouse

There are two types of fogging systems for greenhouse using (1) High pressure foggers (Fog flash through pin atomizer) and (2) Centripetal foggers (direct pressure swirl jet atomizers).

The high-pressure fogger is fed with very high-pressure water, ideally from 500 to 1000 psi. This high pressure water passes through specially designed internal structure of the fogger and finally ejects fog from the orifice in the form of super-fine droplets of water averaging 10 micron size. Due to the very small size of droplets it will evaporate fast, hence cooling and humidifying the air more quickly. These types of nozzles typically put out 5 to 8 liters of water per hour and are generally spaced 2 meter (6 feet) apart.

The centripetal fogger for ventilating high humidity is a self-contained unit incorporating a large fan that forces a stream of air through water ejected from a rapidly rotating nozzle. The water is atomized into an average 30+ micron droplets, which is then forced into a cooling air stream through the greenhouse by a fan attached to the rear of the unit. This system for ventilating high humidity works better when greenhouse is shaded properly and good fan ventilation is present.

Advantage of fogging system

1. Reduction of temperature up to 30° F can be achieved in hot & dry climate.
2. Reduce stress on plants by eliminating the excess heat.
3. Improve plant growth by increasing humidity.

4. Humidity and temperature levels are uniform throughout the greenhouse and precisely controlled, which is not possible by other methods.

5. Requirement of water is less in comparison.

Disadvantage of fogging system

1. Not much effective in hot and high humid climate

2. High initial and maintenance cost.

3. Good quality water is essential.

4. Automation is must.

(e) Fan-pad cooling system: This effective cooling system is useful for large and closeable greenhouses. It is much effective in dry and hot climate and serves dual purposes of reducing the temperature and increasing the humidity simultaneously. However, in high humid situation it will not work properly.

Basically two methods of fan-pad cooling system are available commercially; l) positive pressure system and 2) negative pressure system. (Fig. 43)

The positive pressure systems consist of both fan and pad in the same side of the greenhouse. The fan sucks air from outside and it blows through the wet pad into the greenhouse. Thus the pressure inside the greenhouse is higher than outside. This system is suitable for smaller greenhouses. Steady and uniform spreading of cool air in larger greenhouses by this system is not possible. In order to achieve optimal cooling, the greenhouse may be shaded. This type of cooling system is available in the market in readymade form (as a unit) having a specific capacity (area/volume). For cooling a greenhouse, number of such unit is determined in accordance to the area of that greenhouse.

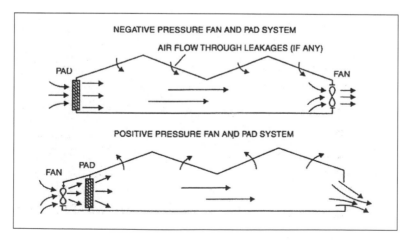

Fig. 43. Negative and positive pressure fan-and pad-cooling systems

The negative pressure systems consist of fans at one side and a pad on the opposite side. Actually the fan and wet-pad are placed in opposite sidewalls of greenhouse along the length. With the fan on, it sucks and throw out the inside air to bring down the inside air-pressure. Due to the inside pressure deficit the hot and dry outside air enters into the greenhouse through the porous wet pad. Water present in the pad is evaporated by the hot airflow and the heat (latent) is removed from the air making it sufficiently cool. This cool air flows from pad end towards the fan reducing the temperature of the greenhouse uniformly.

In the process of movement the cool air from pad to fan again gets heated and finally the fan drives out the warm air outside. The process is continued and the temperature of the greenhouse reduces gradually to the desired level and is maintained as long as the fan is on. Still, there is a temperature difference between fan and pad. For larger greenhouse this system has to be customized for producing proper efficiency. In this system suspended dust particles present outside have a tendency to get into the pad hence regular cleaning is necessary to maintain the air flow rate.

Designing of a fan-pad cooling system: For designing a fan-pad cooling system, a few important information is necessary in respect of the site and the greenhouse. These information are (1) altitude of the site, (2) interior light intensity, (3) temperature difference between fan and pad (in case of negative pressure system), and (4) distance between fan and pad.

In general the temperature difference between fan and pad is restricted within 7°F (4°C) and if it increases further the velocity of air movement by fan is required to be increased accordingly. On the other hand the distance between fan and pad is normally maintained between 100 to 200 feet. More than 200 feet requires more elaborate and expensive equipment and less than 100 feet lower the cross sectional velocity of air movement, which often develops a clammy feeling.

Two main factors shall be considered while designing the fan-pad system of a greenhouse: The ventilation rate or 'Air-removal-rate' (ARR) required for the greenhouse, which finally determines (a) the size and capacity of fans and (b) the area of pad, which cools down the air passing through it.

Hence, to design the fan-pad cooling system for a specific greenhouse the following steps has to be taken.

Gerbera in fan-pad greenhouse

Step-1: First the total air removal from a greenhouse is calculated with the help of 'standard air-removal-rate' (9 cfm/sft or 170 m³/hr/m³). This can be done in the following ways.

Total ARR of greenhouse = Length x Width x Standard ARR.

Now to adjust or correct the above total air-removal-rate, it is required to multiply the Figure by the larger of the two factors e.g. Distance between fan and pad (Factor-D) and House (Factor-H).

The 'Factor-H' is derived as follows.

Factor-H = Factor-A x Factor-L x Factor-T (See Table 20)

All these correction factors (A= altitude, L= light intensity and T= temperature difference between fan and pad) are given in the Table 20.

Table 20: Different factors to standardize the fan-pad system of a greenhouse

Table for altitude		Table for light intensity		Table for fan-pad, temp, variation		Table for fan-pad distance	
Altitude (Feet)	Factor-A	Foot Candles	Factor-L	Temp. diff' F	Factor-F	Distance (feet)	Factor-D
>1000	1.00	4000	0.80	10	0.7	20	2.24
1000	1.04	4500	0.90	9	0.78	25	2.00
2000	1.08	5000	1.00	8	0.88	30	1.83
3000	1.12	5500	1.10	7	1.00	35	1.69
4000	1.16	6000	1.20	6	1.17	50	1.41
5000	1.20	6500	1.30	5	1.40	60	1.29
6000	1.25	7000	1.40	4	1.75	70	1.20
7000	1.30	7500	1.50			75	1.16
8000	1.35	8000	1.60			80	1.12
	8500	1.70				85	1.08
	9000	1.80				90	1.05
						95	1.02
						100	1.00

The tables developed by NGMA, New York, USA.
Note: These factors are to make adjustments in the rate of air removal of a greenhouse standardized below 1000 feet altitude at 5000 fc light intensity and at 7°F temperature difference between fan and pad placed 100 ft apart(Factor-D), which is normally 9cfm/sft or 170m²/hr/m³.

So the actual air-flow rate or ARR of a greenhouse or the total capacity of the exhaust fans will be -

Actual ARR = total ARR x Factor-H (or Factor-D).

Step-2: Next step is finalization of capacity and number of exhaust fans. To determine the collective capacity required of fans, the static air pressure figure of the pad and outside has to be taken into consideration. This is due to the resistance the fan meets while drawing air through the pad and also driving it out against the outside air. Collective capacity shall at least be equal to the actual air-removal-rate. The fans should not be spaced more than 25 to 30 ft. (7.5 to 10m) apart. So on the basis of size (length) of a greenhouse the number of fans can be determined. Now the capacity of each fan can be finalized by dividing the actual-ARR (in cfm or m³/hr) by the number of fans. Then the fans of that capacity can be chosen from the market. The whole process can be expressed in the following way:

Length of greenhouse / spacing of the fans = number of fans.

Actual air-removal-rate / Number of fans = Individual capacity of fan.

The individual capacity shall be adjusted with assumed static pressure. The individual capacity figure will finally determine the size of the fan from the standardized table (Table - 18).

Step-3: Determination of the area of the pad. However, the quality and type of pad should be chosen before determining the area of pad.

The pads are nothing but a thick (4 to 6 inch) porous screen having very large surface area with good wetting property without much retention that allows passing of both water and air through it. Water is applied on the top of the pad, which runs down through it and collects below. Hot air from outside passes through the thickness of the pad and enters the greenhouse after transferring the heat to that flowing water.

There are excelsior pads made from aspen, which are required to be replaced every one or two year. Pads composed of a special corrugated cellulose paper impregnated with insoluble antiriot salts, rigidifying structures and wetting agents are also available. It has life expectancy of ten years or more and is available in different thickness having better efficiency. A comparative statement of different characters of various pads is given in Table 21.

Table 21: Specifications of different types of pad used for fan-pad cooling system.

Pad type	Air flow required (exchange/min)	cfm/min of air flow accommodated per sft. pad	Water flow rate gallon/min/ ft length of pad	Water loss rate gallon/ min/ft/1000 cfm airflow	cfm air flow Pump capacity gallon/sft of pad
4" Aspen	1 to 1.5	150	0.30	0.50	0.5
4" Cellulose	1 to 1.5	250	0.50	0.05	0.75
6" cellulose	1 to 1.5	350	0.75	0.05	1.00

In general, the thickness of pad ranges from 10 to 20 cm (4 to 8 inch). However, the pad area shall match the air exchange/flow rate of an individual fan, which determines the actual pad area of a greenhouse. Generally for every 150 cft of air removal from a greenhouse minimum 1 sft of pad area shall be required. A sample (Table 22) is given to assess the fan pad relationship of different fan size and power and of the 4" pad, made of Aspen.

Table 22 : Air delivery rate and matching pad-area for various size of fan

Fan size (Inch dia.)	Horse Power	CFM in 0.1 inch static pressure	Pad (aspen 4") per fan in sft.
24	0.25	4700	30
24	0.33	5700	36
24	0.5	6500	42
30	0.33	7400	47
30	0.5	8800	56
30	0.67	10200	65
36	0.5	10500	67
36	0.67	12600	81
36	1.0	14200	91
42	0.5	12500	80
42	0.67	15000	96
42	1.0	16800	108
48	0.67	17800	115
48	1.0	19600	126
54	1.0	22800	152
54	1.5	26850	179

Data from Acme Engineering and manufacturing Co., Muskogee, Oklahoma.

Placement of pad: The pad may extend through the entire length of the middle portion of the sidewall. Frequently the pad is mounted on the curtain wall, which is opposite to the wall where the fans are placed. The necessary height of the pad is determined by dividing the total area of the pad by the length and the fans shall be fitted in the same height. However, a few necessary modifications depending upon the type of structure of the greenhouse is sometimes required.

Watering arrangement for the pad: The water is poured on the top of the pad from a perforated (small holes existing 4 inches or 10 cm apart) pipe running horizontally along the top of the pad. The watering is done in such a way that the entire length of the pad gets thoroughly wet. The waterfront gradually moves downward and finally collects at the gutter placed at the bottom of the pad. The collected water is then stored in a reservoir and from there again supplied to the top of the pad by a pump (Fig.44). The water is continually recycled throughout the operational time. On an average 5 liters of water /min /meter of pad length (1/3 gallon/min/foot of pad) is required irrespective of height of pad. The total requirement of water per cycle determines the size of reservoir and as a thumb rule it normally measured @ 1.5 gallons for a foot length of a pad.

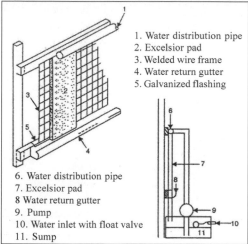

1. Water distribution pipe
2. Excelsior pad
3. Welded wire frame
4. Water return gutter
5. Galvanized flashing

6. Water distribution pipe
7. Excelsior pad
8 Water return gutter
9. Pump
10. Water inlet with float valve
11. Sump

Fig. 44: Diagram of (a) the component and (b) water distribution system of a model fan-pad cooling system

Some additional points to be considered for efficient fan-pad cooling system:

1. Place the fan at leeward side of the greenhouse.
2. If fan exhausts into the windward side, then its capacity shall be increased by 10% or more depending upon the speed of the wind.
3. Fans from one greenhouse should not exhaust warm, moist air toward the pads of other greenhouse unless it is located at least 15m or 50 feet away.
4. When the fans are placed in adjacent walls of two greenhouses, which are spaced within 4.5 m (15 feet), they should alternate their position to avoid face-to-face situation. Actually the position of gables shall be alternated between the two greenhouses.
5. The maximum fan to pad distance should be 30 to 40m (100 to 130 ft.)
6. The pump for watering pads should start first and after complete wetting of pads (2 minutes) the fans will be switched on. This should be done to avoid the clogging of pads by sand and dust flown into the pad from outside.
7. In order to control the water flow to the pad, valves with gauge system may be included in the pipeline.
8. On an average 20 gallon of water per foot length of pad per hour (300 iit./m/hr) is required for this system which is about 7 to 8 times of water loss from the greenhouse due to evapotranspiration.

9. The watering system of pad requires adding of fresh water. The quantity will be determined by subtracting the evaporative loss of water from pad (Table-17) from the water discharged at the top of the pad

Example Problems

1) Fan-pad cooling system: Suppose, a 100ft wide and 300ft long greenhouse is located at 3000ft altitude. Using shade net the highest light intensity is maintained at 6000 fc. The temperature difference between fan and pad is retained at 7°F. They are placed along the length of the greenhouse. The 4-inch 'Cellulose' pad is used and the standard air-removal- rate (ARR) is 8cfm/sft floor area of the greenhouse.

Now, the calculations for developing a fan-pad system for this greenhouse are as follows:

1. The standard ARR of the greenhouse = 100 X 300 X 8 = 240000 cfm.
2. Correction factor for the greenhouse (Facto-H) = Factor-A X Factor-L X Factor-T = 1.12 X 1.2 X 1.0 = 1.34 (Table-20)

Note: Factor-H is greater than Factor-D (1.0) [from table-20].

3. Actual ARR of the greenhouse = 240000 X 1.34 = 321600 cfm.
4. The fans are placed 25 feet apart, so number of fan = 300/25 = 12
5. Now the capacity of each fan = 321600/12 = 26800 cfm.
6. The fan size and power will be 54 inch and 1 Vj hp (Table-22).
7. The area of pad (4"cellulose) = Air flow capacity of fan / Air flow capacity of pad per sft. = 26800/250 = 107.2 sft (Table-21)
8. The total pad area = 107.2 X 12 = 1286.4 sft.
9. So the height of pad — 1286.4/300 = 4.288 feet.
10. Capacity of pump to supply the water to pad = 1/3 X 300 = 100 gallon/ min.
11. The minimum size of water tank for this purpose = 1.5 X 300 = 450 gallon.

2) Forced (fan-tube) ventilation system: Suppose, in the same greenhouse, as above, the side walls are open and covered with IP-net. We want to introduce a fan-tube exhaust system with exhaust fans fitted with perforated plastic tube to maintain an air exchange rate of about 4 cfm/sft floor area, while maintaining the temperature difference of I5°F between inside and outside. Now for that fan-tube system the number and size of the fans may be calculated in the following manner.

1. Standard ARR of the greenhouse = 100 X 300 X 4 = 120000 cfm.
2. Greenhouse factor (Factor-H) = 1.12 X 1.2 X 1.0= 1.34

3. Actual ARR of the greenhouse = 120000 X 1.34 =160800 cfm.

4. Number of fan required = 300/25 = 12.

5. Capacity of each fan = 160800/12 = 13400 cfm.

6. So the fan size & power will be 36 inch and 1 hp respectively.

Note: As the sidewalls of the greenhouse are considered to be open and there is almost no resistance of inflow of outside air through IP-net, a lower standard ARR may be considered here to the extent of 1-2 cfm/sft. However this concept is needed to be standardized.

6.2.2. Temperature increasing system or Heating of Greenhouse

The day-time heat generated in greenhouse due to 'greenhouse effect' is generally conserved to provide heat to the crop at night when the night temperature goes below the optimum level for plant growth. But it is not always sufficient when the ambient temperature is very low. It that situation, additional heating arrangement is required to sustain the high inside temperature, throughout the night, in favour of crop growth.

Thus, when the outside as well as inside night temperature of a greenhouse is much below the optimal temperature required for plant growth, heating of greenhouse is necessary. Besides, in some areas, the night temperature frequently goes well below thus heating is required to avoid frosting. In general the heat required for greenhouse is measured by British Thermal Unit (Btu).

BOX - 14

British Thermal Unit (Btu or BTU): This traditional unit of heat is defined as the amount of heat required for raising one pound of water one-degree Fahrenheit (1°F). In case to express larger Btu, horsepower is used. I boiler horsepower = 33478 Btu.

Commonly heat is now known to be equivalent to energy, for which the SI unit is the joule; one BTU is about 1055 joules. The unit BTU is still important in many fields like pricing of Natural Gas, etc.

Such heating of greenhouse is adding of heat at the rate at which it is lost from the greenhouse. A greenhouse covered by single layer PE-film loses 1.15 Btu of heat through each square foot of covering in every hour when the outside temperature is 1°F lower than inside.

Mode of heat loss: Heat can be lost in three ways e.g. by conduction, by infiltration and by radiation.

Most of the heat is lost by conduction through covering materials of a greenhouse. The amount of heat loss through glass, PVC, FRP, PE film (single layer) and PE-film (double-layer) is about 1.13, 0.92, 1.00, 1.15 and 0.70 Btu/sft/hr/°F respectively.

The second mode of heat loss is by way of infiltration though cracks, or perforations and leakage in the joints around junctions, ventilators, doors etc.

The third mode of heat loss is by long-wave radiation of heat energy from warm objects inside the greenhouse, mostly the steel elements of the structure. The rigid covering materials do not readily permit the passage of radiant heat, whereas PE-film permits the same to a fair extent depending upon the thickness and quality of the film. However, a layer of condensed water film or double layer provides good barrier to loss of radiant heat.

To calculate the heat loss of a greenhouse the following method is followed:

1. Calculation of total heat loss from different surface area of a greenhouse (roof, sidewall, curtain wall). The heat loss figures of these areas, covered by any specific covering material (glass, PE-film etc) are available from standard table.
2. The figure is then multiplied with 'Climatic factors' (temperature and wind velocity), available also from standard table.
3. The result is then multiplied with 'Greenhouse factors' (for roof & sidewall) and 'curtain-wall factor' (for type of curtain wall). Thus the actual heat loss figure is arrived.

6.2.2.1. Different system of heating of greenhouse

Boiler and heater are the two basic heating equipments used in commercial greenhouses throughout the world. However, the procedure of heat transfer and distribution is a different aspect altogether. Apart from these two established heating systems some local, low cost, and temporary arrangement of heating may also be practiced on an emergency basis. This is like arrangements of temporary fire-place inside at night for a restricted period.

(a) Boiler

The boiler is generally used for heating a number of greenhouses or greenhouse range at a time. This system of heating is called 'control heating' system. Here the steam or hot water produced by the boiler is piped to the various greenhouse locations. Hot water systems have been used for small ranges (less than 20,000s ft) and steam is often used for larger ranges. Both these systems have advantages and disadvantages.

The hot water has been customarily supplied at a temperature of 180°F (85°C) in the greenhouse. On the other hand steam is usually supplied at a temperature at 215°F (102°C). The distribution of both hot water and steam is done through pipes to the greenhouse. The size and placement of pipe is designed accordingly.

(b) Heater

Use of heater is the other method where a number of heaters are required for small greenhouse range or big stand-alone greenhouse. Individual heater heats a specific area of the greenhouse i.e. where it is placed. Thus this system is called 'localized heating system'. Heaters usually produce hot air, which heats the respective portion of the greenhouse.

Three types of heaters are generally used in greenhouses viz. (I) unit or forced air heater. (2) convection or simple heaters and (3) radiant or infra-red heaters.

(1) Unit or forced air heaters have three functional parts. One fire-box to produce heat by combusting fuel. It is the initial function of this system. The second function is to transfer the heat through the exhaust system to raise the inside temperature of air with the help of a fan. The fan draws the cool air of the greenhouse and flows the same through the heater.

Thus air of the greenhouse gets heated. This is a continuous process that increases the temperature gradually. Horizontal distribution of warm airflow from this heater is a better system than that of vertical airflow system. The properly designed horizontal unit heater systems can perform all the functions of heating, cooling and forced ventilation with some modification. Heating with unit heater system is cheaper than that of boiler system.

From unit heater the warm air is emitted directly, particularly into the smaller size standalone greenhouses. In large greenhouses a polythene tube of 18 to 30 inch diameter, depending on the length of the greenhouse, with round holes of 2 to 3 inch diameter every few feet along the tube, is installed along the length of the greenhouse to carry and discharge the warm air supplied by the unit heater evenly. If the fan of this heater collects cooler air from outside after switching off the fire-box, then the entire system may be used as a fan tube ventilation system for cooling the greenhouse.

(2) The convection heaters are used in hobby or very small commercial greenhouses owing to its low initial cost. Here, the heat generated in 'firebox' transfer the heat directly from long exhaust pipes into the greenhouse. The pipes are placed along the ground either between beds or beneath the growing benches.

(3) The radiant or infra-red heaters of solar heating systems are low energy heaters that reduce the fuel bill to a good extent (up to 50%). These heaters are placed overhead. The infrared radiation when passes through or touches any object, the electromagnetic energy is immediately converted to heat, thus heating the object without heating the air. In comparison to electrical heater it reduces about 75% of electricity bill. It has three functional parts: (1) collector of heat from solar radiation (2) storage facility to store that heat available only at daytime (3) heat exchanger to transfer the heat, through it, to greenhouse (Fig. 45).

Solar-heating system is a recently developed system to heat a greenhouse and a point of interest to greenhouse owners.

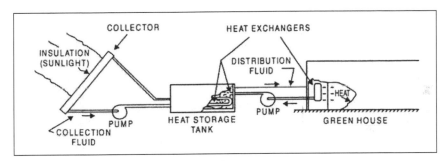

Fig. 45: Sketch of a typical solar heating system for greenhouse

(c) Low-cost temporary heating arrangement using fire-place

Alongside these two established and organized heating system a low cost, crude, and temporary (short duration) arrangement of heating, like putting a standard or local fire-place inside for a restricted period, may warm the greenhouse in a chilling night. A necessary smokestack outlet may also be arranged if it is required to run for a longer period, i.e. more than half an hour.

There are many climatic situations where only for a few days of winter the night temperature goes below the threshold value for plant growth and can destroy the crop. In such places this arrangement is useful and viable in comparison to installing a costly permanent heating system. However, it can be standardized as per situation.

The timing and period of heating through such controlled fire is needed to be calculated properly to achieve the goal. The heat thus generated by the fire/fire-place will warm the micro-climate inside and remain trapped partially for the rest of the night.

Proper design of this temporary heating system can also help the crop to increase photosynthesis by way of injecting CO_2 inside. This technology do not require any energy, thus can be practiced by small farmers of remote places.

Exhaust of heating system and greenhouse: It is important that in all greenhouse heating systems the exhaust shall not come in contact with the crop. The un-burnt fuel, ethylene gas, sulfur dioxide gas etc. present in the exhaust of any fire-box can damage the crop to a great extent. So after transmission of heat from the exhaust, it should be disposed entirely outside the greenhouse with caution to prevent re-entering.

Control over the heating system: The heating system operates mostly automatically by using the thermostat. The thermostats should be placed at the height of growing plant in a location with typical average temperature of the greenhouse and in a light-reflecting box.

Computerized programme may be installed to operate the heating system of a greenhouse.

6.2.2.2. Integrated cooling and heating system

When the day and night temperature vary widely, particularly when evenings with clear sky are cold after a bright sunny day leading to a very low night temperature both the cooling and heating arrangements are often required for a greenhouse.

The fan-tube ventilation (cooling) system discussed earlier [see 6.2.1.1.(ii)(b)], with a small addition and modification can serve as a day-time-cooling cum night-time-heating system. For the said modification, a unit heater system can be attached with the inside of the blower and before the polyethylene tube (see Fig.46). Along with it two louvers are placed with the fan that allow to suck air either from outside to cool the greenhouse or from inside to transfer heat into it before circulation.

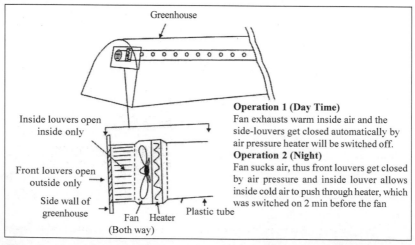

Fig. 46: Sketch of Integrated cooling and heating system for greenhouse

During summer and around summer, in many subtropical highland areas with clear sky the night temperature often falls close to zero. In that situation the above system, for cooling purpose, will not work properly particularly in daytime. So, fogging system, as evaporative cooling, with unit heater and fan (both way) along with perforated PE-tube system shall be the best way to do it. It may be provided in a greenhouse operating system through a single multistage thermostat. It may be considered to be a completely integrated cooling and heating system of greenhouse (Fig. 46).

6.2.2.3. Backup heater or generator system

Power failure in the periods of coolest winter (sub zero temperature) or in the hottest summer (+ 40° C) may invite total crop loss. In snow falling areas except for manually fired convection heaters and proper management of long wave radiation at night, all other heating systems requires electricity. In boiler system it is used for backing up the distribution system. In case of solar heating system it is for running the backup heater. Kerosene or LPG fed manually operated fire-boxes may be used for such situation in a small greenhouse. The simple kerosene Firebox can raise temperature of 12000 cft of air from 25° to 30°F and is considered adequate for providing emergency heat for a greenhouse with floor area up to 1500 sft. These heaters burn 1/2 to 1 gallon of kerosene per hour.

In case of big units a generator is essential not only for power failure in greenhouses with heating unit but also for greenhouses with cooling units. Without fan the cooling systems, other than naturally ventilated system, fails to reduce the inside temperature in hot summer day. So a standby electrical generator is essential for any greenhouse operation. On an average, one-kilowatt capacity generator is required per 1000 sft floor area of greenhouse where continuous heating or cooling or both systems are installed.

6.3. Humidity Control Mechanisms

Problems in Control of Humidity inside greenhouse

In greenhouse climate controlling system, nothing is so complicated as the control of humidity, particularly the high-humidity: Greenhouse air humidity is determined by various factors and controlling of those is virtually impossible. In a greenhouse one can only adjust air humidity in favour of plant/crop and try to avoid extremes.

In greenhouse, transpiration and evaporation takes place from plants and floor of the greenhouse. Some of this moisture, added to the greenhouse air, goes outside through ventilation and/or leakage. With sensible modification of inside heat, the greenhouse can modify the relative humidity to some extent. However, to maintain desired relative humidity in greenhouse, efforts are made to use humidifier (in case of low humidity) or de-humidifier (in case of high humidity) system.

6.3.1. Problem and control of low-humidity

In hot dry area it is very difficult to increase the RH in a sustainable way. Normally 10- 12% RH can be raised up to 30 to 40% in greenhouse condition, if not the inside temperature reduces. The main problem is to get enough water vapour in the air to produce 50% or more relative humidity inside the greenhouse, particularly with drip irrigation system without making everything in the greenhouse soggy wet. However, evaporative-cooling system is the best solution in this situation.

Sometimes during winter when sensible heat is added to raise the greenhouse night air temperature, the relative humidity level might fall below this limit. In that situation it may be needed to operate humidifier or apply water in mist or other form.

To control the low air humidity in the greenhouse, addition of water vapour into the greenhouse air is the only solution. This can be done easily by introducing evaporative cooling system which can take care both the high temperature (which reduces RH) and low humidity problems.

However, the probable solutions of low-humidity are as follows:

(a)   Automatic misting system may be an option, but it is a costly system and has a partial effect on RH. However, a low-cost manual or mechanical misting system with ½ gallon to 3 gallon or 2 to 6 liter per hour flow rate can be adopted in greenhouse in hot daytime to increase humidity. It also reduces temperature to an extent till the level of humidity is low.

(b)   High-pressure fogging is a more costly but more effective system that can reduce temperature and increase humidity in greenhouse. This fogging system can produce 10 micron droplets that quickly evaporate into the air increasing the absolute humidity of the greenhouse without wetting. This will reduce the temperature of the greenhouse to a great extent, increasing the relative humidity considerably. This is called "humidification". In suitable situations this system of humidification can also reduce temperature up to 15°C.

(c)   Fan-pad cooling system can reduce temperature and subsequently increase both absolute and relative humidity to a good extent. However, this system provides uneven cooling of the greenhouse in very hot & dry climate and cannot increase relative humidity sufficiently and homogeneously.

(d)   Water application, through sprinkler/mister or manually, can be adopted inside the greenhouse, if the crop can withstand wetting and have sufficient disease control mechanisms. This method can increases the humidity but with low efficiency.

6.3.2. Problem and control of high-humidity

High humidity prohibits transpiration, particularly with low solar radiation in winter or in rainy season. High humidity increases the feeling temperature to a great extent due to 'heat index' and hampers the growth of the crop. High humidity also increases infestation of disease and pest in greenhouse crop.

In humid or wet season, when the outside or ambient humidity is very high, the inside air of greenhouse (with crop) faces a serious problem of higher humidity. Furthermore, dehumidification is often a problem and there is no amenable to simple solution.

(a) Ventilation and air movement is the best and moderately effective way to control high humidity. Ventilation means exchange of air between inside and outside of a greenhouse. This will effectively reduce both the excess humidity and temperature produced inside by way of replacing humid & warmer air with drier and cooler air of outside.

If the outside humidity is higher, the design is not proper and there is poor natural wind flow, the natural ventilation sometimes does not work properly. In this situation the adverse effect of high humidity, on crop, can be avoided by using an exhaust system along with 'air circulating fans' (for detail see chapter- 6.2.1.1.(ii)).

Exhaust (normal or fan-tube) system may be used to expel the warmer and humid air out of greenhouse rapidly. This creates the desired air circulation in greenhouse that facilitates the natural evaporative cooling and can reduce the inside humidity effectively (for details see chapter 6.2.1.1.(i)).

Though ventilation conflicts with CO_2 enrichment, it is more important to control the high humidity in favour of plant growth.

(b) Heating is a more complicated process than venting to reduce humidity. In an enclosed space, if constant amount of water vapour is heated, the absolute humidity does not change, but relative humidity drops as because warm air can contain more water vapour. However, in greenhouse, filled up with crop, water vapour is not constant due to evapotranspiration, particularly transpiration. So, the overall effect of heating on RH is not easy to predict.

(c) Automatic System of venting and heating: Humidity can only be controlled properly with computer controlled venting and heating systems. It can be designed with the help of well-calibrated humidity and temperature sensors along with automatic vent/exhaust and heating operating system.

(d) De-humidification: It is a process where the water vapour present inside air is forced to condense and drained out of a closed greenhouse. Actually it resembles the methods of using cooling coils in a refrigeration system. Principally a mechanical system called 'de-humidifier' sucks the humid air of greenhouse and cools the air abruptly so that the air temperature drops below the dew-point immediately. This results in condensation of water-vapour thus reducing the humidity of air inside the de-humidifier. This dehumidified air is released again into the greenhouse. If this cyclic process continues inside the greenhouse, humidity will reduce significantly. Generation of heat is the only problem apart from its cost factor. However, in naturally ventilated greenhouse this system will not work, particularly with higher ambient humidity.

(e) Mulching: In hot humid area mulching of bed with plastic can reduce high indoor humidity indirectly by avoiding excessive evaporation of irrigation water from soil. In the process it also reciprocates the addition of water vapour through transpiration into the air, thus reducing the evapotranspiration inside the greenhouse.

6.4. Gas Control Mechanism in Greenhouse

Why gas control? In a greenhouse gas control means control of Carbon dioxide (CO_2) in air. In normal atmosphere CO_2 content is about 0.03 percent, which is equivalent to 300 ppm. However, it can vary from 200 to 400 ppm in micro-climates. Plants consume CO_2 from air at the time of photosynthesis, which yields carbohydrate and other photosynthates necessary for growth of plant. At the same time it liberates oxygen (O_2) into the air. In respect of gas exchange respiration is just the opposite.

Now, in greenhouse the amount of CO_2 is in its lowest state at the afternoon due to continuous photosynthesis, throughout the day, by the high density crop canopy. At night the plant release CO_2 continuously at the time of respiration, thus at morning the concentration of CO_2 will be at its peak (sometime as high as 1000ppm). In day time, as the CO_2 status reduces gradually with time, the rate of photosynthesis reduces accordingly, and will be at its lowest point at afternoon.

At the same time we all know that increase of CO_2 status will increase rate of photosynthesis of plant and here we need the necessary CO_2 management in greenhouse condition, which otherwise is not possible in open field. It has been reported that elevated CO_2 status not only increases the photosynthesis but also increase the water intake by the plant. As a result yield is increased significantly.

6.4.1. Relation between CO_2 level in air and photosynthesis

Photosynthesis: CO_2 + Water (from soil) + Energy (from sun light) $\rightarrow$ Carbohydrate + O_2

Respiration: O_2 + Carbohydrate $\rightarrow$ CO_2 + Water + Energy

Energy generated through respiration is used by the plant directly for its growth and for various functions of growth.

In closed greenhouses, during daytime, CO_2 level can drop to sub-optimal level limiting photosynthesis. In general CO_2 level of 300 ppm is sufficient to support plant growth, but most plant has the capacity to utilize greater concentration of CO_2, attaining more rapid growth. On the other hand CO_2 level below 125-ppm may stop the growth of the crop.

CO_2 uptake by green leaf in presence of light is a common measure of photosynthesis rate. Plant may survive at zero CO_2 levels as it liberates respiration gas (CO_2), which is utilized by that plant to grow/survive. However, the growth is minimal at this situation, and as plant continues to respire in dark, actual loss of weight occurs. The general relationship among CO_2 concentration in air, light intensity and photosynthesis have been worked out in 1933 by Hoover et. Al. (Figure 47)

In closed greenhouse CO_2 concentration may go up to 1000 ppm due to overnight respiration. As sunlight becomes available, CO_2 level starts depleting steadily and goes even below 300 ppm much before noon, (Fig: 48).

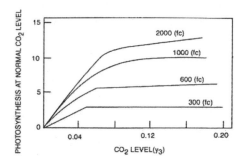

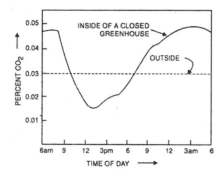

Fig. 47: Relationship between CO_2 concentration, light intensity and photosynthesis in wheat crop (Hoover *et al.*, 1933)

Fig. 48: Diurnal variation of CO_2 concentration in greenhouse (Chandan, 1992)

6.4.2. CO_2 injection

CO_2 injection is also known as "CO_2 Fertilization". Practically CO_2 injection performs two functions in a greenhouse. First it replenishes the reduced CO_2 level of the greenhouse and maintains the normal level of concentration i.e. 300 ppm. Second, according to the crop, additional CO_2 is injected to induce the growth and yield of the crop cultivated inside greenhouse, sometime significantly.

In closed greenhouse, generally found in cooler regions, up to 1000 ppm CO_2 can be injected in daytime depending upon the crop. The crop responses may vary according to the extent to which higher CO_2 level can be maintained.

It has been reported that CO_2 injection in Rose crop @ 1000 ppm has resulted in increase in production up to 53 percent, by way of decrease in the number of blind shoots and increase in stem-length, weight, number of petals etc. This CO_2 injection also reduces the cropping time in winter. Chrysanthemum yields more in the form of thicker stems and greater height when CO_2 is injected into the greenhouse. It may reduce the normal cropping time of 12 to 16 weeks up to 2 weeks by way of increasing the height of the plant. Carnation yield also increases up to 38 percent by CO_2 injection. The time of flowering is also reduced up to 2 weeks through this process. Many other flower crops like Orchids, Snapdragon, Iris, etc response positively to CO_2 injection.

Greenhouse vegetables also have been benefited by CO_2 injection. For example Lettuce crop has been produced in 20 percent less time, while Tomato yield has been increased more than 50 percent. In case of bell-pepper with treatments that received water volumes equal or greater than the evapotranspiration rate, the greatest total fresh fruit mass was observed at the 600 ppm of CO_2 environment.

6.4.2.1. Methods of CO_2 injection/enrichment

Two important methods of CO_2 enrichment process are available for greenhouse cultivation.

1. Combustion: Combustion of hydrocarbons like LPG, paraffin oil, kerosene etc produces CO_2 along with some other gases. A LPG gas burner can maintain a CO_2 level of 1500 ppm in a 450m^2 greenhouse. Kerosene (1.360 m^3/m^3) is able to produce more CO_2 in comparison to LPG (0.785 m^3/m^3) and other fuels.

This combustion method can be practiced only with 99.5 to 100 % pure fuel and only when heating of greenhouse is required.

However, at the point of use, propane has lower carbon content than gasoline, diesel, heavy fuel oil, or ethanol. Natural gas (methane) generates fewer carbon dioxide (CO2) emissions per Btu than propane, but natural gas is chemically stable when released into the air and produces a global warming effect 25 times that of carbon dioxide. With propane's short lifetime in the atmosphere and low carbon content, it is advantageous from a climate change perspective in comparison to other fuels in many applications.

2. Liquid and solid CO_2: Carbon dioxide gas can be liquefied under high pressure or solidified (dry ice) under high pressure and very low temperature. CO_2 gas from this more condensed form is released at low pressure through 3-6 mm diameter fine plastic tube having needle point holes with the help of a set of regulating valves. A good air circulation system is required for even distribution of CO_2 in the greenhouse.

3. Biological emission of CO_2: This is a new idea of enrichment of CO_2, suitable for relatively smaller or hobby greenhouses. These types of systems are a great option if you would like to see what CO_2 can do for your greenhouse on a budget and without a lot of equipment (i.e. timers, regulators and monitors). Here CO_2 injection or addition can be done through fermentation or decomposition process, which can be a good and cheap option for such greenhouse. There are some other specific methods where naturally CO_2 is produced out of any biological reaction or activity. However, it is still at a conceptual stage and some work has to be done to make it a commercial product.

One of such model is 'Exhale bag', which is filled with mycelium (non-fruiting mushrooms) that gives off steady amounts of CO_2 as a byproduct of their metabolism. These bags cost less and can save you a ton of hassles compared to the infamous fermentation or dry ice or any such methods. After hanging your 'Exhale bag' above the canopy of your crop inside the greenhouse you can use a CO_2 test kit to see how much it raises CO_2 levels.

6.4.2.2. Some important factors of CO_2 injection/fertilization

1. Application of fertilizer or rate of fertigation shall be increased with CO_2 injection otherwise crop will face nutrition shortage.

2. Low light intensity reduces the beneficial effect of CO_2 injection. Thus, it shall be done only when light intensity is sufficient or by using supplement light during daytime.

3. Higher daytime temperature for crops fertilized with CO_2 has been generally beneficial. An increase in temperature by as much as 6°C (10° F) is recommended for Roses while fertilized with CO_2. However, in general 3 to 6°C rise, above recommended temperature in daytime, is recommended for CO_2 injection.

4. CO_2 is notably heavier than air, so it is essential that CO_2 is dispensed from above the plant canopy. Oscillating fans in the grow space, particularly around the CO_2 dispenser; will help distribute the CO_2 around the area.

In sub-tropical areas, where closable ventilator greenhouses are used, CO_2 can be injected for only an hour or so each day in the morning when the ventilators remain closed. After that, the ventilators are opened or the cooling fans are turned on, which neutralizes the effect of CO_2 injection.

In greenhouses with permanently open ventilators the problem in reduction of CO_2 level below 300 ppm is absent as the incoming air replenish the CO_2 deficit inside. However, due to the same reason it is also not possible to increase the CO_2 level inside the greenhouse above 300 ppm by any means.

6.4.2.3. CO_2 Injector

It is the CO_2 generator or cylinder with pressure gauge, that can provide 1500 ppm CO_2 in a 5000-sft greenhouse. High light intensity and to some extent excess heat is helpful to increase the efficacy of the injector.

However, a proper system of injector involves using a CO_2 sensor or simple timer, a CO_2 regulator, and either a CO_2 tank/cylinder (available for purchase or refill) or a CO_2 generator. Tanks are typically the better option for smaller greenhouses and for ease of use. Whereas CO_2 generators tend to be the better option for larger spaces since they run off of propane and can generate large volumes of CO_2 at low cost.

The timer is set to turn the regulator on for 15 minutes after every 90 minutes or so, depending upon the requirement.

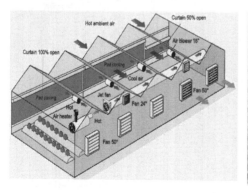

Sketch of fan-pad cooling system for greenhouse   Cellulose cooling pad for greenhouse

Fan of a fan-pad cooling system

Carbon dioxide generator

4 way fogger

Fan-tube cooling system

Fan-tube ventilation system

Air circulating fan

LED light for plant

Different greenhouse management instruments

7

Basics of Greenhouse Cultivation Technology

7.1. Introduction

Cultivation of crop is one of the most important aspects on which the success of a greenhouse depends. Principally it is similar to that of open field cultivation; but it is a completely different proposition in comparison. In greenhouse it needs (1) high precision technology, (2) special management, and (3) high quality inputs. A special crop husbandry is extremely necessary, integrating all these three aspects, for successful greenhouse cultivation.

Why special crop husbandry? – Ancestors of each and every crop of modern time once grew out of its own in wild, without any human intervention. Gradually it has been domesticated or brought under cultivation by progressive alteration of its natural growing and reproduction process. Even the genetic nature also altered gradually. Till recent past these development of domestication process has been restricted principally on reproduction, planting, nutrient supply, water supply, and plant protection aspects. Interestingly most of these aspects are still very much supplementary in nature, as a good percent of it is still dependent on nature and its growing system. Actually, we cannot confer any significant change, physically or genetically, in their adaptation against the natural elements like rain, sunlight, temperature, humidity, gas supply, etc. However, we try our level best with a compromising note on local climate.

Hence there is ample scope to go for complete alteration of their growing habit/ practices in relation to each and every aspect of the climatic elements. It is a very difficult job, because, for a small change in any individual aspect of plant growth, all the other aspects are affected. Thus, one has to make a proper and holistic balance of alteration of all the growth aspects for each and every individual domesticated plant, we call as crop.

So, when we introduce a climate control system to the crop, in the form of greenhouse, we have to change all other aspects of growing need, like nutrient, water, canopy management, etc in a far more precise way. For this reason we have to introduce a completely new brand of technologies and management practices for every individual crop in accordance to the greenhouse-technology adapted for it. It should be complementary, inclusive, and complete process and can be called Greenhouse cultivation technology.

The greenhouse cultivation technology has been compelled to follow a different crop husbandry practice, compared to open field, because of reasons mentioned below.

1. Economic return from cultivation of crop or from activities like nursery etc should justify the high initial expenditure.

2. For cultivation of crops, high yield, both qualitative and quantitative, is required to ensure high net return.

3. The desired level of production, out of the cultivar's genetic capacity, is possible only when the specific husbandry practices and controlling the adverse effect of climate are adopted.

4. Apart from high precision agronomic practices, introduction of some specific package of practices (training and pruning) to accommodate more and effective bio-mass per unit of land area is important. Increase the height of canopy in comparison to width or pulling up the crops irrespective of climbing or creeping habit is helpful in this aspect.

5. Controlling of natural phenomenon always have some adverse effects on the microenvironment where it is imposed. In case of greenhouse it obviously relates to the crop grown under it, thus, more precise and innovated crop husbandry is required to counter these. This has already been discussed in the previous chapters.

To have a basic idea about the 'greenhouse crop husbandry' the following aspects are to be considered properly.

(a) Selection of crop and variety

(b) Raising of healthy and disease-free seedling or plant material

(b) Preparation of bed or root media and planting

(c) Installation of irrigation and fertigation system

(d) Adoption of proper plant protection measures

(e) Execution of suitable Canopy management procedure

The following text describes in brief the above mentioned aspects in general.

7.2. Selection of Crop

The selection of greenhouse technology is very much dependent on the crop or plant to be grown under it. This is the 2nd important aspect after climatic factors, which have to be considered seriously to implement greenhouse technology. The basic purpose for adoption of greenhouse technology is to control the climatic factors in favour of optimal growth of the crop or plant decided to be grown under the structure.

There are two basic purposes for growing plants under a greenhouse: (1) crop production and (2) Rearing or production of plant materials i.e. Nursery or Conservatory (conservation greenhouse).

For nursery, selection of a specific crop is difficult. In general, a range of plant materials are grown or reared in the greenhouse. The technology of greenhouse for nursery is very much different from that of crop production. Furthermore, the need of greenhouse for nursery or conservatory is limited in respect of cultivation. Basically it is a supplemental activity for crop production. Thus, our prime goal is to specify the greenhouse technology for production of crop, and for that matter the selection of crop is discussed in detail below.

7.2.1. Basis of Selection

As the greenhouse cultivation is much costlier than open field cultivation, crops with the following criteria have to be selected:

1. Crops with capacity to Increase the yield manifold, which is otherwise not possible in open condition.
2. Crops that have the capacity to respond to canopy-management for accommodating more bio-mass per unit land area of bed/land.
3. A specific crop that cannot be produced in open climatic condition of that area.
4. Crops capable of Off-season production in greenhouse.
5. Crops with extended growing season
6. Crops with Socio economic demand and good market prospect

Based on above six important aspects, selection of crop for greenhouse cultivation can be considered. These aspects are guided by combined output of some information, which is narrated below.

7.2.1.1. Points for Selection of type of crop (Flower, Vegetable, fruit, etc.)

1. **Market:** Information regarding indigenous & export market, post-harvest facility, transport etc, are the few points to be looked into before selecting the crop and its corresponding variety.
2. **Differential capacity to increase yield in relation to open field cultivation**: It will depend on the availability of standard technology for a specific crop in a specific climate. Frequently we find that some

vegetables show 5 to 10 times more yield when grown in greenhouse condition with improved crop husbandry practices.

3. **Off-season production:** Flower, vegetables, and fruit, which are produced seasonally, has a great demand in off-season or festive season.

4. **Status of the available climate control mechanism:** The greenhouse and its climate control mechanism available frequently determine the selection of type of crop. For example, flowers have a clear advantage in greenhouse over vegetables in respect of requirement of pollinating agents, which is not easily available in greenhouse. On the other hand vegetables require less sophistication in respect of climatic control mechanisms as flowers are more climate sensitive than vegetables.

7.2.1.2. Points for selection of Groups (Genus) of the selected type of crop

1. **Micro-Climate:** The outside climate, which directly affects the output of climate control mechanism or inside micro-climate of a greenhouse, will determine the species of flowers or vegetables to be selected for cultivation. For example Rose and Carnation among flowers, Cucumber and Lettuce among vegetables, and Strawberry and Melon among fruits require completely different type of micro-climate.

2. **Physiological potentiality to increase yield**: Some species have natural ability to extend its potentiality to increase the production and some have not. For example, any climbing or vine type crops like Cucumber, Tomato etc have a clear edge over the single head or root crops like cauliflower, radish etc in respect of increase in the yield potentiality from unit area through proper crop husbandry. Thus, in vegetable, *cucurbitaceae* and *Solanacea* crops have a clear edge over other genus of crops for greenhouse cultivation.

7.2.1.3. Selection of variety (cultivars) of selected crop

1. **Genetic capacity**: It is an important aspect to produce better quality and higher quantity in greenhouse situation. For example, in cases of cucumber and tomato, the parthenocarpic and indeterminate varieties respectively, are suitable for greenhouse cultivation. This type of variety is essential to cope with lack of pollinating agent in greenhouse and increase yield in a small piece of land under cover.

2. **Consumers demand:** The choice of colour, size, formation, etc of a specific variety changes with time, which can be accommodated by proper selection of variety of any particular species of crop.

7.2.2. History and present situation of selection of crop

History of greenhouse says that, till today, generally a limited number of floricultural-crops, vegetables, and a few fruit-crops have been grown under greenhouse condition. However, till recent past, flower was the first choice for greenhouse throughout the world. That is why specific greenhouse technology for specific flower crop has already been evolved and practiced. However, the scenario is changing now-a-days, and vegetable under greenhouse is spreading very fast.

In recent years few specific vegetables are being grown under greenhouse in many countries on commercial basis. Off-season production and very high yield are the basic two reasons to grow such vegetable in greenhouse. However, quality of produce and better pest management are also the other aspects of spreading of greenhouse cultivation of vegetables.

As some vegetables are produced through cross-pollination and as the greenhouse restricts the pollinating agents from entering into it, the selected varieties of such vegetable crops should either be self-pollinated or be 'parthenocarpic' (Cucunber) in nature. For self-pollination also, often specific pollinating agents (bumble bee) or other mechanisms (hand or vibrator) are required to enhance pollination.

BOX - 15

Parthenocarpy: Under certain conditions, unfertilized egg develops into a seedless fruit and this phenomenon is called as 'parthenocarpy'. In general fruit development begins with pollination and continues with seed development; so parthenocarpy is an exceptional situation where normal sexual process is omitted.

The plant, which have the ability to develop seedless fruits are termed as 'Parthenocarpic'. Two types of parthenocarpy are recognized in plants e.g. **(1) Vegetative parthenocarpy** and **(2) Stimulative parthenocarpy**. The first type takes place in species like pear or fig or cavendish banana, where fruit develops naturally without pollination. The natural function of the fruit, in this case, is help to protect the varietal purity. The second type takes place only after pollination but does not require fertilization or seed setting for continued growth of fruit. Some grape species can form seedless fruit through stimulative parthenocarpy.

Parthenocarpy is frequently desirable in crops where seedless fruit is preferred by the consumer, such as cucumber, watermelon, grape, tomato etc. In case of greenhouse cultivation, where pollinating agents are barred from

---

entry, parthenocarpy is the best solution for cross-pollinated crops like cucumber etc. So the horticulturists deliberately apply certain growth regulators/hormones to induce parthenocarpic fruits. Auxin and gibberellin are generally sprayed to produce parthenocarpic tomato and grape respectively.

In case of monoecious plants like cucumber and squash, constant femaleness should be maintained in the plant for bearing the fruit and for that matter plants may be treated with cytokinin or ethylene. Morphactin (chlorofluorenol) may be used to induce artificial parthenocarpic fruit setting in some plants like cucumber.

The researchers are still trying hard to evolve specific greenhouse technology for crops of other choice and of different climatic situation, which will be gradually introduced for commercial cultivation.

1. **Flower Crops**: Rose, Carnation, Gerbera, Anthurium, Chrysanthemum, Orchids, Lilies etc are the main flower crops generally grown in greenhouse. In general perennial flower crops are chosen for greenhouse production.

   The selection of variety of the flower crops is a vital aspect. The viability of greenhouse flower cultivation depends on the market price. The market price depends on the specific demand of the market in respect of quality and standard of the flower. Specific variety should be identified for cultivation depending on the market demand.

   Most of these flowers have great demand all over the world throughout the year. However, in some specific season/period, like marriage, religious & social festival like Christmas, Valentine-day etc, the demand rises to its peak. To maintain quality of flower the climate control mechanisms are generally used with very high precision

2. **Vegetable Crops**: Tomato, Sweet/Bell pepper, Cucumber, Lettuce, and Egg-plant are the main vegetable crops grown under greenhouse condition. However, other crops like Leek, Squash, Melons, Gourds, Peas, Beans etc are also considered suitable, and are grown in greenhouse throughout the world.

3. **Fruit crops:** Strawberry, Musk melon, Blueberries, grapes (table var.), etc are common fruits generally cultivated under greenhouse condition.

In India, the greenhouse cultivation of vegetables augments the reduction of cultivable land in and around cities and fulfils the market demand of different vegetables throughout the year. 100 to 200 kilometers from any metro and 50 to 100 kilometers from any big city is the best area for greenhouse cultivation of quality vegetables. Greenhouse technology has tremendous potential to increase the yield of selected vegetables, which can be 3 to 4 times of open cultivation and also extends the production season. The quality of crop would also be better than the crop produced in the open field.

Greenhouse technology provides scope to produce organic or toxicity free vegetables that have high demand in the market, particularly in respect of salad vegetables.

7.3. Seedling Production and Management

The following points are the basic reason of discussing seedling production and rearing of tissue-culture plants in nursery (small specially designed greenhouse) separately with much impetus.

- Very high cost of quality seeds (Hybrid) and plant materials (tissue-culture),
- Very low immunity and subsequent susceptibility of seedling/plant towards biotic stress like disease and pest infestation,
- Off-season production of seedling or plant produced in laboratory need much more protection from all abiotic stresses, thus need acclimatization before transplant in the bed of greenhouse,
- Very healthy seedling/plant have no competition for nutrient and have a pronounce root system leading to best possible growth.
- Continues growth (no time loss) while the seedling/plants are re-established in beds of greenhouse.
- It gives uniformity of crop and assures the expected yield,

Now, one can understand why this technology is so important for cultivation of crop in a greenhouse. For any crop in any climate this is an important aspect to do and to manage accordingly.

In general the smaller greenhouses of 100 m^2 or more are used for commercial purposes as nursery or conservatory. For any small individual producer 50 m^2 or less (even 20m^2) is sufficient for this purpose. Basically every greenhouse should have a link with such a nursery.

7.3.1. Nursery for seedling production

For vegetable production under greenhouse, regular supply of disease-free seedling is needed, which is essential particularly for high cost vegetable seeds. This seedling may be grown in disinfected soils of a raised bed or in containers or trays having large number of cells filled up with treated soil-less media (coco peat).

Now-a-days use of such tray, called 'pro-tray' or 'plug-tray', is considered to be the best method of seedling production, particularly for vegetable seedling.

Some vegetables like cucurbits and sweet corn require special care during nursery raising and transplanting. Such trays contain cells or cavities of specific shape and size.

Shape of cells may be like inverted pyramid, round or hexagonal. The standard cell size of a tray available in the industry is 2.5 x 2.5 cm and the number is 100 cells per tray. In general one seed per cell is recommended. Deeper and bigger cell (5x5 or 6.5x6.5 cm), having 50 or 35 cells per tray, is required for large size seeds and to promote faster growth. Seedlings raised in larger cells have been reported to give higher early yield compared to the smaller cells but the overall yield remains same.

Table 23: Dimensions of cell and size of different pro-trays

Cavity	Tray Sizein mm	U.Dia x L.Dia.in mm	Depthin mm
104	300 x 485	32 x 20	38
98	265 x 525	32 x 20	38
50	272 x 525	48 x 32	50
70	360 x 555	48 x 32	55
38	300 x 485	55 x 38	115
9	300 x 300	80 x 80 sq.	80
209	290 x 495	25 x 18	28

Artificial soil-less media is used in plastic pro-trays for raising healthy and vigorous seedlings. For this, the following procedure can be followed.

1) Peat-moss or coco-peat (coarse) are generally used as root-medium along with vermiculite and perlite in a ratio of 3: 1: 1 respectively. Only peat-moss or coco-peat can be used for this purpose.

2) The cells of the tray should be filled and compacted with the root medium, like coco-peat properly. Then it is soaked by sprinkling of water.

3) The seed is put in the cells, one seed per cell, and gently pressed with index-fingure. A thin layer of dry root media is applied over it and the tray is placed in the dark after covering with a plastic sheet.

4) After 3-4 days of sowing, the trays are to be uncovered and placed inside the nursery-greenhouse to protect the emerging seedlings from rain and other aberrant climatic conditions. This will help in even germination.

5) Water and nutrient has to be supplied from time to time, and gradually the seedlings will be ready for transplanting (25 to 35 days after sowing of seed).

Greenhouse also provides a favourable climatic condition for proper germination of the seeds. Germination of different crops, particularly the vegetables seeds, require different climatic conditions (Table -21).

Procedure of healthy and disease free Seedling production in pro-tray

Seedling production unit in greenhouse made up of wood

Table 24 : Temperature (°C) range for germination of seed and growing of transplant.

Crop	Germination temperature	Dayrequire	Growing day temperature	Growing night temperature
Tomato, Eggplant	21-24	3-4	18-21	12-18
S. pepper Chilli	26-28	4-6	18-21	12-18
Cole crops	18-24	2-3	10-18	8-14
Most Cucurbits	24-30	2-3	21-24	12-18
Onion	18-24	3-4	16-18	8-15

Source : Indo-Israel project, IARI, Pusa, New Delhi.

Excess stem elongation of seedlings due to low light, high heat, over fertilization and watering is a common problem of seedling production. Controlling the above mentioned factors or application of growth retardants like Cycocel @5-10 ppm depending upon the crop may reduce stem elongation. Uniformity in growth may be achieved by shifting positions of tray to avoid variation of air circulation and application of water homogeneously.

Depending on the crop and climatic condition, the vegetable transplants become ready in pro-trays in 25 to 32 days after sowing. The coco-peat base root-medium can hold sufficient water to transport the seedlings up to a 24 hour journey.

7.3.2. Hardening of seedling or tissue-culture plants in nursery (conservatory)

Before transplanting of plant material into bed/pot, particularly the tissue-culture plants, hardening is essential for reducing transplanting shock and sudden exposure to completely different environment from a highly controlled one (laboratory etc).

Actually the term "hardening" refers to any kind of treatment that results in firming or hardening of plant tissue to fight against the new environment. Such treatments reduce the growth rate (temporarily), thickens the cuticle and waxy layers, reduces the percentage of freezable water in the plant and often results in a pink color in stems, leaf veins and petioles. Such plants often have smaller and darker green leaves than non-hardened plants. Hardening results in an increased level of carbohydrates in the plant permitting a more rapid root development than in non-hardened plants.

Any of the following can be used to harden transplants. A combination of all or a few of these techniques at one time is more effective.

1. **Reduction of water**: water lightly at less frequent intervals but does not allow the plants to wilt severely.

2. **Temperature:** Expose plants to lower temperature than is reported as optimal for their growth. Note: Plants could be placed in a proper greenhouse/nursery that does not allow the extreme ambient temperature during day and night.

3. **Fertilizer:** Do not fertilize, particularly with nitrogen immediately before or during the hardening process. A starter solution or liquid fertilizer could however be applied to the hardened transplants one or two days prior to transplanting into the garden or at the time of transplanting.

4. **Exposure to sunlight:** Gradually expose the plants to more sunlight. This results in the development of a thicker cuticle layer thereby reducing water loss.

This also provides a better and quick crop stand. This gradual exposure of the transplant to the anticipated growing conditions of the crop is the key of proper hardening. The process hardens or toughens the transplants by way of acclimatizing.

By Slowing down the growth rate by way of gradual reduction in supply of water and nutrients and withdrawal of shade, the acclimatization and hardening is done. This will provide strength to the transplants so that they can withstand shortage or excess of temperature, humidity, water, and to some extent resist the pest attack.

Interestingly, after completion of the process when the plants are exposed in normal growing condition in the greenhouse, initially it grows faster than its usual rate. Gradually the growth rate settles down.

7.4. Preparation of Root Media for Bed, Pot, Trough, etc.

Generally crops are cultivated in a greenhouse either in beds or in pots or in trough and bag of different kinds. Pot culture has some specific advantage in monitoring of individual plant and to manage the spreading of disease. However, pot culture is costlier and it has been reported that the vigour, particularly the height of crop, is relatively less in comparison to bed culture when treated similarly. Thus, if different crop husbandry is not followed for pot, then the bed culture for greenhouse cultivation is a more acceptable method and commonly practiced around.

Root medium: Before preparation of bed or pot mixture, one should have a clear idea about proper function of the root medium that provides the crop an optimal growing condition. A root medium must have the capacity to provide four basic functions to support good plant growth:

1. Media must provide anchorage or support to the plant. Individual roots grow among soil particles and provide a firm foundation for physically supporting the stem as it grows.

2. Media must serve as a reservoir for plant nutrients. These elements must be in an available form, in sufficient quantities, and in proper balance for adequate growth of plant. With the exception of carbon and oxygen, plants obtain all essential elements from the growing medium.

3. Media must hold and provide available water to the plant needed for one irrigation or rain to the next.

4. Media must provide adequate gas exchange (supply O_2 and remove CO_2) between the roots and the atmosphere. Respiration is required by roots to provide the energy for uptake of water and nutrients and root growth.

Individual media components can provide some or all four of the functions of a medium but not at the required levels of each. Thus, it is very difficult to get or prepare such an ideal growing media, which naturally or inherently have all the above-mentioned characters. Most of them may be good in few activities and poor in other.

However, this concept may be used to convert a growing medium into a high-quality growing media by developing the poor functional qualities with outside help. For example, sand is a good medium in respect of providing sufficient air space to plant but poor in other aspects. Water is necessary for supplying nutrient but excess water induces poor gas exchange and anchorage.

For good quality cultivation, mixture of few such items in proper proportion can provide all the facilities a crop requires for its optimal growth.

However, when only water is considered as growing medium then it is termed as 'hydroponics'. It has been discussed separately later on.

7.4.1. Soil as a root media

Soil, by nature, is a mixture of sand, silt, clay particles and organic matters that have the ability to do all the functions mentioned above. However, optimization of all the functions cannot be assured by any kind of soil. The proportionate mixture of sand, silt, clay and organic matter is the key factor for a soil to act as a very good root media. The problem of optimizing the ability of soil is that the functions stated above (for ideal root media) are inter-related. For example more air space reduce the water holding capacity and *vice-versa*. However sandy loam soil, where sand, silt, clay and organic matter are present proportionately, is conducive to root growth and may be considered the best type of soil for plant growth.

1. Bed preparetion for gerbera in eastern India

2. Construction of bed for Orchid

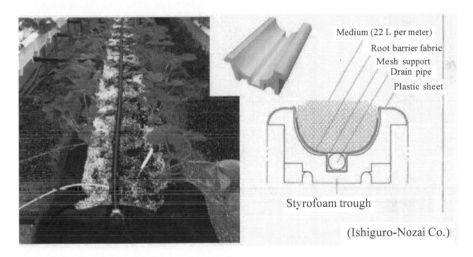

Medium (22 L per meter)
Root barrier fabric
Mesh support
Drain pipe
Plastic sheet

Styrofoam trough

(Ishiguro-Nozai Co.)

Styrofoam trough system (Ishiguro Nozai Co.) Substrate volume: 22 L per meter
Substrate volume per plant: 2.4 L (>2 L recommended); Composition: 50% Perlite+25% Coco
coir+25% Peat (Air porosity: 15.6%@inch column

Strawberry on troughs in greenhouse

Gerbera bed preparation and sterilization

Bench prepared for Dendrobium Orchid

Small hydroponic vertical farming

Large hydroponic vertical Farming

To obtain maximum growth of crop in greenhouse-condition, the root medium should be prepared carefully and specifically. There are two types of root medium: (i) soil-based root medium and (ii) soil-less root medium.

7.4.2. Soil-based root media

Most of the greenhouses for crop production use soil-based media. Traditionally, the soil- based medium is composed of equal parts by volume of loam field soil, coarse sand, and well- decomposed organic matter adjusted to proper PH level.

Sandy field soil is compensated by an increase in organic matter and decrease in sand, while clay soils require more sand.

A moist mixture of equal parts of loam soil, sand and organic matter, weighing about 40.82 kg per cubic foot, is suitable for bed preparation. However, parts of the volume of well-decomposed organic matter may be replaced with coarse organic material like coconut husk, coarse bark, peat moss etc that increases aeration. This type of adjustment may be done to create more balance among the required characters of root medium like water holding capacity, aeration, steady nutrient supply etc.

Maintenance of soil-based medium: Decomposition results in loss of organic matter of the medium. So periodic application of organic matter should be done customarily, it can be done each year when the root medium is pasteurized or churned on routine basis.

Organic matter, coarse or fine, at the rate 5 to 10 percent of the volume of root media, depending upon nutritional status and drainage status proves to be good.

The medium, prepared with loam soil containing more clay particle and organic matter, may show drainage problem in due course. Cracking on drying is the symptom of this condition. This problem is remedied by addition of coarse sand to this media at the rate of 10 to 15 percent of the volume of the medium as a single dose.

7.4.3. Soil-less root media

The first attempts to standardize greenhouse mixes in the U.S. were by researchers at the University of California in the 1950's. "The UC mixes" were composed largely of different combination of sand and peat moss. A few growers had problems with these mixes because of inconsistent sand supplies. The first truly standardized "soil-less" mixes were the "peat-lite" mixes developed at Cornell University in the 1960's. These media consisted of various combinations of peat and perlite or vermiculite and these are still in use today. 'Cornell Mix A' is half peat and half vermiculite while 'Cornell Mix B' is half peat and half perlite. Most of the soil-less root media, in use today, is variation on the Cornell mixes. Lime is frequently mixed with it to correct the pH and to supply calcium and magnesium.

In initial stages of greenhouse cultivation, particularly in crop specific greenhouse, soilless root medium is popular for the following reasons:

1. Proper field soil is not available, and transportation of soil is not feasible.
2. Components are readily available in standard and uniform quality.
3. Consistent chemical and physical properties for best plant growth.
4. Generally, do not require pasteurization.
5. Cost of preparation and handling is competitive.
6. Base fertilizer is low but adequate to begin plant growth.
7. Perfect for application of automation in the greenhouse.
8. Preparation can be mechanized easily.
9. Mixes are consistent from batch to batch.
10. Mixtures can be light-weight and suitable to transport for a longer time.

Components of soil-less root media: It is actually a mixture prepared artificially, which does not contain any soil but certainly provides better root growth. That mixture shall perform all the functions of an ideal root media, mentioned earlier, in an efficient manner. Pit moss, bark, sawdust, compost, vermiculite, sand, perlite etc are the components, which are mixed in proper proportion to prepare a root medium. At the time of selection of components the character of the individual items should be considered meticulously to create the best possible root media for proper plant growth. The character of individual medium is given in table-25.

Table 25: Characters of different components of root media

Components	Water retention	Nutrient retention	Aeration	Light weight	Used for (Root medium)
Field soil (loam)	Yes	Yes	No	No	Soil Based
Sand (coarse)	No	No	Yes	No	Both
Saw dust (rotted)	Yes	Yes	No	No	Both
Coir pith and piece	Yes	Yes	Yes	Yes	Both
Manure / compost	Yes	Yes	No	No	Both
Vermiculite	Yes	Yes	No	Yes	Soil less
Perlite	No	No	Yes	Yes	Soil less
Bark (3/8 to 3/4 inch)	Yes	Yes	Yes	No	Both
Paddy husk	No	No	Yes	Yes	Soil based

7.4.4. Chemical amendments of root media

Plant nutrients are present in the soil solution in ionic form either as cations, with a positive charge, or as anions, with a negative charge. Component root-medium like clay, peat/organic, and vermiculite have negatively charged surfaces that electrostatically attract and hold cations. Cation exchange capacity (CEC) is a measure of the ability of the media to hold and buffer cations with the soil solution. The cations may be listed in order of binding strength as follows:

$$H^+ > Ca^{+2} > Mg^{+2} > K^+ > NH^+ > Na^+$$

Cations on the left are bound to soil particles (clay, organic, and vermiculite) more tightly than those on the right. As the concentration of cations decrease in the soil solution, forces opposing cation binding cause a release of bound nutrients into the soil solution for root uptake. Conversely, as the concentration of cations increases in the soil solution, most of the exchange sites become saturated. Weakly adsorbed cations (such as Na^+) are readily replaced by stronger ones (like Ca^{+2}) so are more subject to leaching.

The chemical amendment of root medium, particularly the soil-less medium, is an important factor. It is almost impossible to prepare a homogeneous mixture of versatile character like natural soil artificially, in respect of its nutritional status and chemical reactions. Thus, to make the mixture most efficient regarding supply of nutrient to the plants the following chemical amendments should be done.

1. Adjust pH level at 6.2 to 6.8 for general crops and for acid loving crops 5.0 to 5.8 - Use limestone or gypsum to raise pH and use sulfur or leaching process to lower it. The application rate will be calculated properly.

2. Supply of phosphorus should be provided through application of single super-phosphate or liquid phosphatic fertilizer on a proper dose per unit of root-media.

3. A complete range of trace element or micronutrient is imperative as these soil-base and soil-less mixtures commonly tend to show micronutrient deficiency.

4. Both soil-based and soil-less mediums usually contain nitrogen and potassium that can last for 2 to 6 weeks and can only supply nutrient up to the young seedling stages. Then nitrogen and potassium shall be applied in the media at proper doses. Unless the slow release form of nitrogen and potassium is available, it is better to go for an established fertilizer programme like fertigation after planting.

BOX - 16

pH, its problem and solution: pH represents the status of soil reaction and is negative logarithm of hydrogen ion (H^+) concentration. This indicates degree of acidity or alkalinity of soil. A pH reading of 7 is considered as neutral (equilibrium of acidity and alkalinity). The value above and below 7 is considered as alkaline and acidic respectively. Regarding status, pH 4 is 10 times more acidic than pH 5 and 100 times more acidic than pH 6.

Problem: Soil pH regulates nutrient availability for plant. It also affects the biotic (microbial in particular) population of soil. However, pH range of 6 to 8 is good for almost all aspects of soil-plant relationship.

Soil pH may rise due to high evapotranspiration or any other reason that accumulates or increases salts in soil. It may be low due to continuous leaching or washing off the bases from surface soil due to heavy rainfall and/or high rate of drainage.

Solution: Low pH may be corrected by adding limestone ($CaCO_3$) or gypsum ($CaSO_4$) to the soil in proper dose, depending upon the status of soil pH.

High pH may be rectified through leaching or washing off the excess salts by water by increasing soil drainage system. Chemically, sulfur compounds may be added to the soil to lower the pH.

7.4.5. Pasteurization or fumigation of root media

Now-a-days it is an essential practice for all greenhouses. However, in warm conditions, where the inside environment of greenhouse does not freeze, humidity is high, temperature is warmer, the scope of development of disease causing organisms is more. Continuous culture of a single crop, or cycle of a few crops, in a greenhouse provide a continuous host on which disease causing organisms build up. The problem of nematodes is also aggravated due to the reasons stated above.

In early days, to combat against these soil-bome problems in green house, root media were replaced annually. After late 50s this labour consuming and cumbersome process has been gradually replaced by pasteurization or fumigation.

Root medium of a greenhouse is generally pasteurized annually. However, sometimes it is done between every crop. This increase in frequency is occasionally necessitated due to the proliferation of disease-causal organisms in the greenhouse.

Summer is preferred for pasteurization of greenhouse root media because the crop production is usually low, labour is available and root media are warmer.

The pasteurization of root media has many benefits other than control of disease and nematodes. It helps to control soil-borne pests and weeds. Sometimes control of specific pest or weed is a more important reason for pasteurization of root medium in a greenhouse than that of disease and nematodes.

Methods of Pasteurization

Pasteurization may be accomplished by means of injecting steam or any suitable chemicals into the soil. These two methods are described separately.

(i) Steam pasteurization: The root medium should be loosened before pasteurization. This will help the movement of steam though the pores and transmit the heat rapidly within the medium. The root medium should not be dry and if dry, addition of water speeds up the rate of pasteurization. But, excess watering slows down the speed of pasteurization. Moistening of root medium a week or two prior to pasteurization is the best procedure, which breaks the dormancy of many unmanageable weed seeds, and then pasteurization destroys them easily.

Chemical and physical inputs, like super-phosphate, limestone, pitted or coated micro nutrient, inorganic complete fertilizers and the slow releasing fertilizers, are to be added into the root medium prior to pasteurization. All these materials are also pasteurized during pasteurization process of the medium without any adverse or negative effect.

Procedure: The steam for pasteurization is generally produced by boiler of different types or by generator. Separate boiler designed for pasteurization or the steam boiler used for heating of greenhouse or small portable steam generator may be used for pasteurization purposes.

The main job of the steam is to raise the temperature of even the coldest/ farthest point of the root medium, minimum up to 82°C (180°F) and maintain the same for 30 minutes. The steam does not have to be generated under high pressure for pasteurizations. Once it is released in the root medium, it is under very low pressure.

Distribution of steam into the root medium can be done through buried perforated pipes having paired holes (with 1/8"' to 1 /4th inch diameter) on opposite sides spaced in a gap of every six inches. The five inch diameter canvas-hose can also be used for pasteurization of root medium particularly in raised benches. Covering of bed or benches for at least 30 minutes after injection of steam is recommended.

For field or for beds of large greenhouses 'stem rake' may be used to inject steam into the root media.

Problem with steam pasteurization: Overheating, i.e. heating for more than 30 minutes is harmful for the root media. Pasteurization of soil-based root media can create Manganese and ammonium toxicity, which above a certain level, is not good for crop/plant. The organic matter based soil-less root media is also affected by the ammonium toxicity.

(ii) Chemical fumigation or pasteurization: In greenhouse, instead of injection of steam, chemicals like Methyl Bromide, Chloropicrin (tear gas), Formaldehyde, Hydrogen per oxide, etc may be injected or drenched for treatment/pasteurization of the root medium.

Table 26. Detail of different chemicals for soil treatment

Chemical	Type	Application	Toxicity	Period of rest (days)	Use restriction	Home garden use
Chloropicrin	Compressed gas	injection	High-poison	14-30	Yes	No
Dazomet	Granules	surface	moderate	21-30	No	Yes
Formaldehyde	Liquid	drench	moderate	21-30	No	Yes
Metum sodium	Liquid	drench	moderate	14-21	No	Yes
Methyl bromide	Compressed gas	injection	High-poison	14	Yes	No
Vorlex	Liquid	injection	High-poison	7-21	No	No
Hydrogen per oxide	Liquid	Surface/ drench	moderate	1 day	No	No

Procedure of application: Methyl Bromide is an odour-less volatile material, which is hazardous to human health. To warn against exposure of methyl bromide two percent chloropicrin or 'tear gas' is added with Methyl Bromide, as irritant. Methyl bromide is available in smaller cans or larger cylinders for tractor mounting injection systems.

The root medium of a greenhouse, either in pile (for bench or pot) or in beds are to be loose enough, and moist up to the stage of planting and temperature level should be maintained at 50°F (10°C) or higher, before pasteurization. The cans contain one pound of Methyl Bromide, which pasteurizes one cubic meter of root medium. The bed or bench or pile should be covered by a polyethylene sheet for at least 24 hours after injection is started.

The can is generally put on an applicator and is placed at the top of the bed or bench. The applicator helps to drive a hollow spike into the can allowing the chemical to escape through the tube into the medium below the cover. The exposure should continue for at least 24 hours at a temperature of 60°F (I5°C) and higher. Methyl Bromide can also be applied by tractor for larger greenhouses.

Chloropicrin or Tear gas is also used as fumigant, but its poor penetration into the plant tissue in the root media is a disadvantage. However, it is used at the rate of 3 cubic centimeters (cc) per square feet of bench or bed and for bulk media the rate is 3 to 5 cc/cft. It can be done by hand injector having a spike at the base, which is pushed up to one foot into the medium and this releases 3cc of liquid tear gas (chloropicrin) from the end of the spike. After application, the bed or bench or pile is covered with polyethylene cover and the temperature of the media should not go below 60°F (15°C). An exposure time of 1 to 3 days is needed. For larger areas tractor drawn chisels can be used to apply chloropicrin. After completion of pasteurization with chloropicrin the medium should be aerated for seven to ten days before planting.

Formaldehyde is a commonly used chemical to sterilize the root medium. Drenching of root medium with formaldehyde mixed with water, @ 25 ml per liter, is the usual practice. After drenching, the soil or root medium will be covered with plastic Film. It is found very effective to check the serious disease like 'damping-off.

Hydrogen peroxide (H_2O_2) is used for sterilization of growing medium for seed bed and propagation trays. Washed river sand can be used instead of perlite and vermiculite if disinfected with hydrogen peroxide. When soil or bed-mix is made with compost, sand and other additives, chemical solution of Hydrozen peroxide with silver shall be used to destroy soil pathogens, including nematode worms that can harm plants. First the beds are soaked with irrigation water. Then the soil-mix is thoroughly and uniformly soaked with irrigation water mixed with H_2O_2 + silver solution @ 3 to 6 % solution (30 to 60 ml/lit) depending upon the requirement. The application rate of the solution will be 1 lit/sqm. There is no need of soil covering and only 6-12 hours will be allowed for the complete decomposition of the chemical, preferably overnight, then crop can be planted.

(iii) Solarization: The soil or root medium can be disinfected in warmer climate by covering soil with transparent plastic in hot summer days. This will increase the heat of the soil to a great extent and destroy many soil borne pathogens and insects. However, solarization coupled with fumigation works better to control effectively a good number of soil borne insects and diseases.

7.4.6. Hydroponics

It is the method of growing plants using water as root medium. In 1936 W E Gerick first successfully grew a wide variety of crops in water supplemented with all necessary plant nutrients and termed the method as 'hydroponic' or 'water culture'.

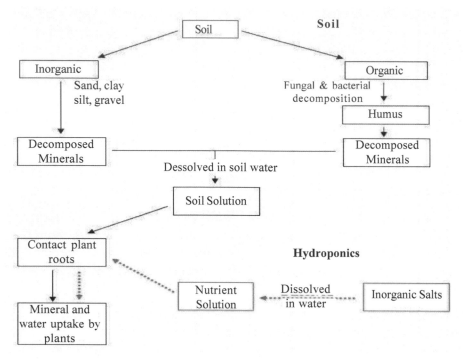

Fig. 49: Origin of essential elements in soil and hydroponics

It is basically a type of soil-less culture, and is a subset of hydroculture, which is the art of growing plants in soil-less medium (dependent only on water for every requirement of plant). One of the most obvious aspects of hydroponic is to make or choose a soil-less medium and corresponding technique they should use. Different media are appropriate for different growing techniques.

Apart from water, as the main component, different substrates can be used to create a suitable soil-less hydroponic system. In general Coir/Coco-peat, Rice-husk, coarse sand or gravels, parlite, and vermiculite, are used as the substrates of hydroponics. Apart from these materials Rock-wool, Grow-stone (glass-wool), burn-Clay aggregates, small Brick bats, Bark or wood pieces, etc may be used as the other substrates of any hydroponic system. These substrates solely or in combination with 2 or 3 other suitable material can make hydroponic system along with water.

There are different growing techniques used by the growers, which are different in principle of growing and operational aspects. A brief detail of these techniques are given below.

1) **Static culture:** Here the plants are grown in containers or tanks of nutrient solution, even in big jars, plastic buckets, tubs, etc. The solution is usually

gently aerated to add oxygen in to the solution. It may not be aerated, and in that case the solution level is kept low enough that a portion of roots are above the solution (in the air) so they get adequate oxygen. Proper arrangement for standing of crop on water surface or a few centimeters above it should be designed.

2) **Continuous flow culture**: Here the nutrient solution along with dissolved oxygen constantly flows past the roots. It is much easier to automate than the static culture because sampling and adjustments to the temperature and nutrient concentrations can be made in a large storage tank that will supply to many plants at a time.

A popular variation of continuous flow culture is the 'nutrient film technique' or NFT, whereby a very shallow stream of water containing all the dissolved nutrients required for plant growth is re-circulated past the bare roots of plants in a watertight thick root mat, which develops in the bottom of the channel and has an upper surface that, although moist, is in the air.

3) **Aeroponics or fogponic system:** It is a system wherein roots are continuously or intermittently kept in an environment, saturated with fine drops (mist or aerosol) of nutrient solution.

4) **Passive or flood and drain sub-irrigation system:** These are the simplest form of hydroponics, sometimes called as semi-hydroponics, and are not suitable for commercial purpose. Like soil-less culture here the inert porous substrates supply the solution to the plant root.

5) **Different deep-water culture** is also accepted and practiced by the Hydroponic growers around the globe. Here plant roots are exposed (suspended or dipped) in a solution of nutrient-rich, oxygenated water.

7.4.7. Hydroponic Nutrient Solution

Hoagland solution, the most popular and authentic solution, was developed by Hoagland and Arnon in 1938 and revised in 1950. It is one of the most popular solution composition for growing plants. The Hoagland solution provides every nutrient necessary for plant growth and is appropriate for the growth of a large variety of plant species. The solution described by Hoagland and Arnon in 1950 has been modified several times, mainly to add iron-chelates. The original concentrations for each element are shown below.

N 210 ppm, K 235 ppm, Ca 200 ppm, P 31 ppm, S 64 ppm, Mg 48 ppm,

B 0.5 ppm, Fe 1 to 5 ppm, Mn 0.5 ppm, Zn 0.05 ppm, Cu 0.02 ppm, Mo 0.01 ppm.

The Hoagland solution has a lot of N and K so it is very well suited for development of the large plants like tomato and bell pepper, etc. The solution is very good for the growth of plants with lower nutrient demands as well, such as lettuce and aquatic plants with the further dilution of the preparation to 1/4 or 1/5 of the original.

However, there are different source of the nutrient which are as follows.

Table 27: Different source of nutrient to prepare Hydroponic solution

Individual Ingredients	
Ammonium nitrate (NH_4NO_3)	33.5% N
Calcium nitrate liquid (7-0-0-11) 12.1 lb/gal ($Ca(NO_3)_2$)	7%N, 11% Ca
Calcium nitrate ($Ca(NO_3)_2$)-dry	15% N, 19% Ca
Calcium chloride ($CaCl_2$)	36% Ca
Potassium nitrate (KNO_3)	13% N, 36.5 K
Monopotassium phosphate (KH_2PO_4)	23% P, 28% K
Phosphoric acid (H_3PO_4) 13 lb./gal.	23% P
Potassium chloride (KCl) - greenhouse	51% K
Magnesium sulfate ($MgSO_4$)	10% Mg, 14% S
Solubor	20.5% B
Copper sulfate ($CuSO_4$)	25% Cu
Zinc sulfate ($ZnSO_4$)	36% Zn
Iron, Fe 330 chelated iron, etc.	10% Fe
Manganous sulfate ($MnSO_4$)	28% Mn
Sodium molybdate ($Na_2(Mo)_4$) (liquid), (11.4 lb/gal)	17% Mo
Sodium molybdate (dry) $Na_2(MoO_4)$	39.6 % Mo
Soluble Trace Element Mixture (S.T.E.M.):	1.35% B
	7.5% Fe
	8.0% Mn
	0.04% Mo
	4.5% Zn
	3.2% Cu

Several other solutions are commercially available required to prepare the nutrient solution for hydroponics. Crop-wise formulation of nutrient solution for different techniques of hydroponics has also been developed.

Actually, nutrient solution is the key for success of hydroponics. There are basically two methods to supply the fertilizer nutrients to the crop: 1) premixed products, or 2) grower-formulated solutions. The two methods differ in the approach to formulate the fertilizer and the resulting nutrient-use efficiency.

However, ultimately four basic aspects are to be managed technically by a grower for successful implementation of hydroponics, which are;

(1) Root aeration,

(2) Physical support to crop,

(3) Darkness in the root zone, and

(4) Supply of all sorts of plant nutrients and additives.

Root aeration by way of providing air to the roots directly or through water/ nutrient solution is essential to prevent anaerobic respiration and facilitate aerobic respiration.

Physical support is must for the plant to stand, hang or float erectly on or above the nutrient solution.

Root darkness prevents growth of algae and other aquatic flora around the roots, which interferes with the proper function of root.

Supply of plant nutrients is critical than that of field as because water in isolation have no inherent or buffering capacity to supply nutrients to any plant. The detail of it was discussed earlier.

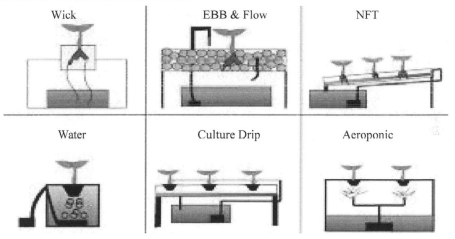

Fig. 50: Models of different type of hydroponic systems

7.4.8. Aquaponics

It is a unique integrated system of farming where aquaculture or fish-culture is done and used for hydroponic crop production without any fertilizer/nutrient application.

It is an interesting and most modern environment friendly farming technique, which is mostly practiced under greenhouse. The crop grown under greenhouse and the fish-tank also be placed inside the greenhouse or may be placed just outside of the greenhouse. However, it is really difficult to standardize an aquaponic system, as because it deals with two alive material (fish and crop) and have to maintain a super-fine coordination among the two.

Basic principle of Aquaponics

- There will be a suitable tank with waist-height water and suitable fish supported by necessary aquaculture practices.
- At the same time there will be a suitable hydroponic system used for necessary crop production, without any water & nutrient supply system.
- The fish-waste, secreted through its gills and excretory system, containing *Ammonia* which is converted into *Nitrite* first and then *Nitrate*. This *Nitrate* enrich the tank-water along with many other organic materials and microbes.
- A certain portion of this tank-water will be applied to the associated hydroponic system, for plants on a regular basis. This Nitrate and other essential materials present in that water will be absorbed by the plant in the growth process, and will yield profusely.
- Through this process, the excess water available from the crop, which are purified by crop, are released again in to the tank and complete the cycle.

The advantages of such a aquaponic system

- The high concentration of fish eliminates weed, mosquito larva and other organism/insects harmful to crops.
- The continuous recycle of nutrient rich water and continuous use by fish and plant reduces all kind of toxic run-off from either aquaponics or aquaculture.
- In Aquaponics the crop only use 10% of water in comparison to soil-base cultivation, and even use less water than hydroponic system.
- Here, one don't need to look after the water and nutrient management aspects of the crop and only focus is on feeding the fish to grow.
- It can be small, to any size and can be put anywhere including indoors, apart from greenhouse.
- Finally it can provide an opportunity to harvest both fish and crop.

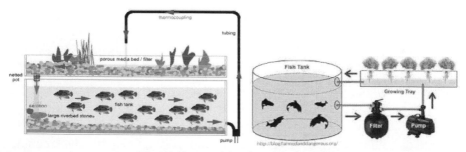

Fig. 51. Two models of aquaponic system

7.5. Irrigation for Greenhouse

In greenhouse, application of water is an essential part as in general, apart from shade house, the greenhouses do not permit rain inside. Naturally the greenhouse plants are completely dependent on irrigation water not only for its growth but also for survival.

As the irrigation is not dependent on nature and attracts expenditure, greenhouse water is given to the crop with proper measurement and through highly efficient methods.

To obtain maximum possible growth of a crop, watering or irrigation in proper quantity and in time is essential. Both under and over watering, even for a very short period, reduce growth of a greenhouse crop.

Effect of under watering: Under watering leads to temporary wilting of plant, i.e. closure of stomata, that retard transpiration and photosynthesis thus reducing growth. It reduces elongation of young developing cells resulting in smaller leaves and shorter stem internodes thus stumping the growth. In extreme cases burning of the margin of leaves or dropping of immature leaves occur.

Effect of over watering: Initially the excessive vegetative growth makes the plant taller and susceptible to wilt under bright sunlight. Further, over watering reduces the oxygen (or air) content of root medium by saturation of soil, resulting in damage of young roots. A damaged root system cannot take up water and nutrient readily thus causing wilting, stunted growth and shows nutrient deficiency symptoms in the plant.

Three types of irrigation methods may be considered for greenhouse cultivation system, which is narrated below.

7.5.1. Hand application of water

Water is directly applied to the growing area of the greenhouse through a measured container or through suitable flexible hose @ 30 lit/minute. This practice is not very common and suitable for hardy and easy-growing crops in shelter type or low-cost/tech or common greenhouse. The main problem of this method is high application rate, which frequently creates saturation of root-zone area and enhance the possibility of disease infestation. Its application efficiency is also low. Apart from this it involves a risk of over or under watering in pockets. Naturally its water-use-efficiency is also poor in comparison to other methods.

7.5.2. Sprinkling or misting of water

Generally sprinkler is not used for greenhouse irrigation purposes because of (i) small growing area and (ii) un-even water application i.e. sprinkling of water in the areas where it is not needed (Paths, walkways etc.). In some specific (Nursery) cases mist-sprinklers or mini sprinklers are used for irrigation.

Misting is used in greenhouses particularly for nurseries and for young plants. Apart from watering it also does the job of humidification, which sometimes is essential for development of young planting materials.

Several types of misting system are available in the market which have the capacity to distribute necessary water in an even and localized manner. This increases the efficiency of irrigation and is very useful for tropical greenhouse.

Modem misting nozzles have multi mode capability like spray mode, localized watering mode, humidification mode and localized atomization mode for cooling. Thus in dry areas this system can take care of seedling establishment, irrigation, humidification and cooling operations of a greenhouse. In case of low plastic tunnels without mulching, misting is a very useful method of irrigation and humidification.

On an average the size of water droplets emerging out of a mister ranges from 50 microns to 100 microns. Mister can be used overhead or below the canopy of the crop. The average rate of discharge is 10 to 20 lit/hr depending upon the purpose and requirement. (See chapter- evaporative cooling)

7.5.3. Drip Irrigation

In commercial greenhouses, it is the most popularly used method of irrigation for production of flowers or vegetables. It is also the best and most efficient method for growing any kind of horticultural crop both in open field and greenhouse.

The method involves of watering the ground surface above the root zone area drop by drop or by trickling. It is 'spot application' of water as per need at a very low rate (2 to I0 liters/hr).

In India, since ancient times, people practice a modification system of drip irrigation for watering house plants in dry periods. In this system a pitcher with a small hole at the bottom, plugged with porous substances is placed over the plant like Basil etc. The water of the pitcher trickles drop by drop from the hole due to gravitational pull and soaks the root zone area of that plant. In present days this method is used to irrigate tree plants in dry areas and called as "pitcher method".

In 1940s, in England, the idea of using plastic pipes and hoses, having minute holes, for trickle system of irrigation for greenhouse plants became feasible and localized irrigation was first used. Modern system of drip irrigation was developed in Israel during 1960s for open field cultivation. An Israeli engineer, Symcha

Blass observed that a tree near a leaking pipe exhibited vigorous growth while compared to others. This led to a different kind of work on irrigation system, which ultimately came to know as drip irrigation system. In the same year Sterling Davis of USDA also successfully attempted sub-surface placement of drip lines on citrus and potatoes.

Advantages of Drip Irrigation System

1. Spot application reduces evaporation and other losses of irrigation water.
2. Only target plants receive water, reducing weed problem.
3. Saves water up to 70% in comparison to other irrigation methods.
4. Low rate (I -5 lit/hr) of application of water moisten only the root zone thus avoiding any kind of percolation or seepage loss.
5. This system of watering maintains an ideal soil-water-air relationship optimally favourable to root system and increase the growth of plant or yield by 10 to 100% while compared to other irrigation methods.
6. Sustainably moistens the root zone but creates a dry micro-climate around the canopy thus reducing incidence of pests and disease.
7. This system of irrigation provides a unique scope to apply fertilizer through irrigation water, called fertigation: this increases the efficiency of fertilizer use and reduces the loss of nutrients supplied through fertilizers up to 40 to 60 percent.

Disadvantages of Drip Irrigation System

1. High initial investment particularly for close spaced crops.
2. The tricky technicality and inadequate technical input.
3. Maintenance, particularly for clogging of drippers.
4. Not suitable for plant population based field crops like cereals, pulses and oil seeds etc.
5. Planning and layout do not provide sufficient scope to increase the area in due course.

7.5.3.1. Component of drip irrigation system

A drip irrigation system consists of six distinct components viz. (1) Power generating segment. (2) Filtration system, (3) Fertilizer/agro-chemical injection segment, (4) Water distribution or pipeline segment (5) Water emitting segment and (6) Control head segments.

1. Power generating arrangements: This arrangement is the starting point of the drip irrigation system. Pressure is required to send water from its source to the ejection or emission point. This shall overcome the head loss occurred in between due to filtration unit and frictions along the distribution pipelines and its bends. Generally requirement of pressure is not very high but for a large system the total pressure requirement is higher and it is provided with pumps of required power. For smaller units gravitational head can be used to create necessary pressure. The job of this pressure head is to supply water up to the delivery point where it will emit drop by drop almost with zero velocity. However, the drip irrigation system operates under pressure as low as 0.15 to 0.2 kg/cm^2 and as high as 1 to 2 kg/cm^2.

2. Filtration System: Proper water filtration is essential for smooth functioning of drip irrigation system. Physical clogging of emitters or drippers can be kept under control through this filtration system.

Effect of clogging- Even a small percentage of emitter clogging can drastically affect the uniformity of water application. With as low as 5 percent emitter clogging, the statistical uniformity can be reduced to approximately 75 percent.

When physical factors (sand, silt, and other suspended materials) are the main reasons of clogging, good quality filters are used in a drip system, operated at 1000 lit/hr/m². However, three basic types of filters are used in drip irrigation system either individually or in combination.

(a)   **Sand filter**: It can effectively remove sand particles and hence used as pre treatment. Suspended other solids may be removed by sand filters (Fig-52).

(b)   **Media filter**: This type of filter consists of fine gravel and sand, placed in a pressurized tank, followed by ring-way valve. It can reduce both sand and suspended materials efficiently.

(c)   **Screen/disk filter**: It is the simplest design, made up mostly of plastic for removing suspended materials effectively. Depending upon the quality of water the filtration system may include a single or combination of screens.

3. Fertilizer and other agro-chemical injection segment: A separate air tight fertilizer mixing tank is installed, which acts as a pressurized vessel with inflow and out flow pipes/tubes connected with main water line having a pressure regulation valve in between two connecting points (fig. 38). Actually a portion of irrigation water collects fertilizer mixture through sucking (ventury) or through pressure difference (mixing tank) or by the help of external force (injector) and return to the main irrigation flow. (The detail is in the fertigation section of this chapter). Other agro-chemicals like pesticide, fungicide, nematicide etc, which are systematic in action, may be applied to the tank, in proper dose, which is finally applied to the crop through irrigation water.

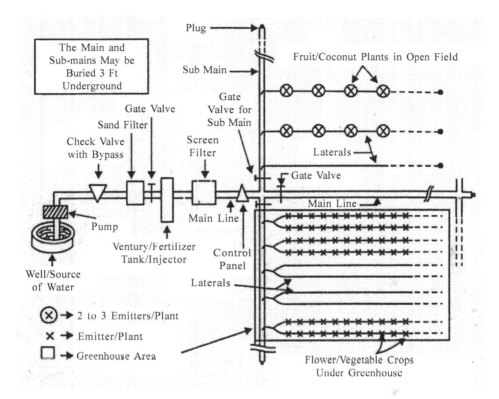

Fig. 52. Sketch of high pressure drip irrigation system and its component for greenhouse

A few 'organo-phosphates' and 'carbamade' insecticides have proved to be moderately successful for systematic control of sucking/piercing insects like thrips, aphids, etc. Many other systematic insecticides have been found effective through drip system against respective pests.

Phytopthora diseases are the main reason for which fungicides are applied through drip irrigation system. Contact fungicide like 'ethazol', 'benomyl' and systemic fungicides like 'ridomil' may be used through drip system to control Phytopthora diseases like 'root rot'.

Applying nematicides through drip irrigation system is rapidly gaining popularity, since it is required in lower doses. Contact nematicides like 'standack', 'vydate', and 'nema cur' are used through drip system.

4. Water distribution or piping system: The distribution line consists of a network of a main line, sub-mains, and/or laterals, generally made up of either 'high density polyethylene' (HDPE) or 'low density polyethylene' (LDPE) 'polyvinyl chloride' (PVC) pipes. The main line is larger in size with a range of 25 to 50 mm diameter, depending upon the size of line and number of emitters. The sub-mains, having diameter of 15 to 35 mm, are generally used for larger plots with different blocks. The buried main and sub-mains are commonly made up of HDPE PVC pipes. However, the laterals are normally made up with LDPE PVC pipes of smaller diameters e.g. 10 to 20 mm.

The appropriate fittings like elbow, tee, cross, plug etc, made up of HDPE PVC or mild steel are available and used to connect the pipes of the distribution system of a drip irrigation system.

5. Water emitting segment: It is the device placed inside (in-line dripper) or outside (on-line dripper) of a small hole of I mm diameter in the lateral lines, which are called as emitter of dripper. These drippers are spaced along the laterals according to the requirement of crop and emit water almost at zero velocity with very low discharge rates i.e. I to 4 liters/hour.

Different types of drippers have been designed and are used nowadays. Constant discharge with specific rate (pressure compensating drippers), capacity to avoid clogging, adaptability to force flashing, suitability of placement, and the cost are the points considered during designing of drippers.

6. Controlling system: The function of drip irrigation 'control system' is to measure, monitor, and regulate the flow of the water and other agro-chemicals running through the drip irrigation system. It may be done through simple valves to most sophisticated computerized control system. However, a drip irrigation system shall at least consist of different kinds of valves, meters, and switches, which work in isolation or in combination in a pre-planned mode.

There are two basic type of controlling system used for greenhouse irrigation; (1) Switch/timer based control system, (2) soil-moisture linked computer control system.

In the first case, an on-off schedule is finalized on a notional basis depending on the soil moisture status and stage of the crop. According to that schedule the on-off time is fixed and operated manually or with the help of a 'timer' with a clock.

In the second case the soil-moisture data generated through a probe, is collected from the greenhouse bed and fed into a computer loaded with necessary software programme. The programme finalizes the schedule of supply of water in respect of time, which is a continuous process running throughout the growth period of crop under greenhouse.

7.5.3.2. Clogging of Drip irrigation system

Proper monitoring during installation and running of a drip irrigation system is necessary to avoid clogging, which imposes one of the major problems of this system. For that matter the following aspects should be taken care of.

(a) Quality of irrigation-water, the basic reason of clogging of the drippers

Analysis of water not only provides the information about its applicability according to crop requirement but also gives an idea about the possible clogging problem in a drip-irrigation system. So it is essential to estimate water quality before making a plan for installation of a drip-irrigation system.

There are three factors associated with irrigation water that causes 90 percent of clogging in a drip irrigation system. Physical, chemical and biological aspects of irrigation water have been rated as 35. 22. and 37 percent responsible for clogging, respectively.

Physical factors of water means the presence of inorganic particles, like sand, silt, and clay, organic materials and microscopic aquatic plants (phyto-plankton) and animals (zooplanktons), which frequently passes through ordinary filters and reach the emitting parts leading to emitter clogging.

Chemical factors, i.e. presence of chemicals/salts that leads to deposition of Calcium Carbonate (CaCO,), Iron-oxides etc in the narrow passages of the emitters and starts clogging.

Biological slimes, filaments, bacteria, algae and other microbial deposition are the most serious factors that frequently clog the drippers. Generally this occurs in presence of iron and hydrogen sulphide.

(b) Solutions against clogging: Prevention and reclamation is the main job to fight against clogging of emitter.

Out of prevention methods (1) selection of proper emitter, (2) proper filtration of irrigation- water, (3) regular field inspection, and (4) scheduled pipe line flushing are the basic four methods, which reduce the clogging of emitters. These are done individually or in combination as per necessity.

Reclamation of partial or fully clogged system may be done by the following methods.

1. **Chlorination**: Addition of chlorine is the primary means to control clogging occurred due to the activity of microbial (algae, fungi, bacteria) materials present in the irrigation water. Calcium-hypo-chlorite, sodium-hypo-chlorite (1-2 mg/lit on continuous basis) and chlorine gas are used to reduce or remove clogging of emitters.

2. **Acid treatment**: The chemical salt deposition, which causes clogging, can be cleaned by acid treatment. A solution of dilute hydrochloric acid (36 %) of 0.5 to 2 % by volume of water, if introduced in laterals for 10 minutes, can effectively remove $CaCO_3$ deposition thus reducing clogging.

3. **Pipe line flushing**: Flushing the main, sub-mains and laterals with pressurized water can remove the potential clogging materials like sand, silt, organic/ biological matters present in the system. The flushing valves shall always be provided at the end of the main streams and manual flushing or some type of lateral flushing device is required occasionally (may be once in every six month) to augment the drain valves. In case of severe clogging, the pipeline flushing may be done by air compressor with a pressure of about 70 meter head of water.

7.5.3.3. Drip irrigation system for small farmers (Gravity flow drip irrigation)

This system utilizes the gravitational force and does not require any external power to operate. It works with low pressure of about 1.5 to 2 meters water height. It can be planned in the following way, both for greenhouse and open field (Fig. 53).

1. A 200 to 1000 liter capacity water tank is mounted on a platform of 1.5 to 2 meter height to store water for irrigation through drippers at the root-zones of the targeted crop. The water can be drawn manually or otherwise and stored in the tank after completion of one irrigation cycle for a specific area designed according to the capacity of the tank. However for a 500 liter tank to irrigate flowers/vegetables the feasible area will be 700 square meter.

2. An outlet pipe of 25 to 40 mm diameter made up of black alkathen pipe (BAP) or polyvinyl chloride (PVC) pipe is fitted as main line at the base of the tank. The diameter of pipe is dependent on the number of emitters i.e. requirement of discharge.

3. A gate valve is provided in the mainline to regulate the flow from the tank. A filter, handmade with fine brass or copper wire of 200 mesh sieve or simple water filter used for domestic purposes, may be fixed just after the gate valve (Fig. 53). If sub-mains are required, a separate gate valve for each sub-main is to be provided to maintain the schedule of irrigation. The end points of the mains and sub-mains are to be closed with plugs. Joints are fitted and if needed clumped properly.

4. As per laid out beds and according to the spacing of crop the laterals (best possible length is about 17 m) are placed on the ground before planting, one end of which shall be connected firmly with main or sub-main pipe line and other end plugged by simply bending the end point 360° and clipped or tying it with a thread. Laterals are generally made up of 8 to 12 mm diameter LDPE pipes having emitters along the length, with a discharge capacity of 0.6 to 1 liters/hr, spaced according to the spacing of the crop.

Planning of this drip irrigation system

The main obstacle of this system is the length of the laterals, which has to be restricted to a certain limit, depending upon the number of emitters. However the following information is essential to plan such a low-pressure drip irrigation system.

- Water requirement of a single plant in liter/day.
- Size of storage tank in liters.
- The number of drippers/plant.
- Time gap for refilling the storage tank in hours.
- The number of plants that can be irrigated by the storage capacity of the tank in one cycle and the size of the land in respect of length of laterals.
- Plan of fertilizer application through this system.

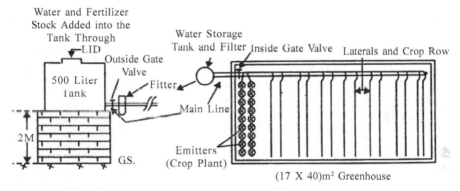

Fig. 53. Low pressure (gravity-flow) drip irrigation system for greenhouse.

Example of low pressure drip irrigation system

1. Tank size - 500 liters, Crops - Tomato, egg plant, cucumber or gerbera. Distribution - Only mainline (25mm ID) and laterals (8-12mm ID) providing one emitter (0.6-1 lit/hr)/plant, Area - average covering area is 600 to 700 m^2, Maximum length - in an average 17-18 meters, Fertigation - Nitrogen and potash is fed through irrigation water.

2. Tank size - 200 liters, Crops – younger fruit plants, Distribution - Mains (38mmID), sub-mains (25mmID) and laterals (12mmID) providing 3 emitters/plant @ 2 lit/hr, Area-0.2 ha,

Gravity-flow drip irrigation system

7.5.3.4. Design of a Drip irrigation system for greenhouse

The design of a drip irrigation system, especially the layout, is done on the basis of water requirement of crop and materials used for the pipeline. The design should indicate the type and number of emitter per plant and pumping schedule.

The steps involved to design a drip irrigation system for a greenhouse and corresponding estimates are as follows.

Step - 1

1. Estimate the command area i.e. the area of greenhouse, by mapping and proper layout of main, sub-mains and laterals. While-ascertaining the area the scope of expansion is to be taken into consideration.

2. Estimation of the number of plant with necessary spacing of the proposed crop or crops of the total command area is required to be finalized.

Step - 2

1. Estimation of peak water requirement, liter per day (Ipd) of the proposed greenhouse crops:

The daily water requirement for a full-grown plant can be calculated as

Water Requirement (WR) = [A x B x C x D x E] + Cwr

Where A = Pan evaporation inside the greenhouse (mm/day),

   B = Pan factor (0.7),

   C = Cropping area/plant (m^2),

   D = Crop factor (For ull-grown it will be I),

   E = Wetted area (Widely spaced =0.3, closely spaced = 0.7).

 Cwr = Crop water requirement (lpd/p!ant)

Example

Suppose the average IMD pan evaporation data of a greenhouse in the month of May (the driest month) is 213.6 mm. So the daily average pan evaporation in that period is 6.83 mm/day.

Now the daily water requirement of a vegetable plant having the spacing of lm x 0.5m is (i.e. WR) = [(6.83 x 0.7 x 0.5 x 1 x 0.7) + Cwr] Ipd/plant.

= [1.67 + Cwr] lpd/plant

2. Estimation of number of emitters and water discharge per emitter per hour:

- The average discharge per emitter available in the market ranges from 1 lph to 8 lph (liter per hour). The standard pressure head required for emitter is considered as 4 m.
- After selection of emitter, depending upon the total WR the number of emitter per plant can be estimated. In greenhouse generally one emitter/ plant is used and only the rate of discharge has to be selected accordingly.
- The performance of emitter is dependent on infiltration rate of soil. For example the emitters of 1, 2 and 3 Iph discharge capacity wetted 0.1, 0.2, and 0.3 m^3 area respectively with 1 cm/h infiltration rate of soil. Increase and decrease of infiltration rate will proportionately decrease and increase the wetted area.

3. Estimation of operation hour per day

Considering the WR and discharge rate of emitter the operation hour/day shall be determined.

Step - 3

1. Estimation of length and diameter of main, sub-mains and laterals

The length of sub-main and laterals can be estimated from the greenhouse layout. However, the diameter of pipes can be estimated considering a) the maximum discharge required and b) estimated frictional head loss in pipes considering the multiple openings in laterals.

2. Calculation of frictional head loss

- The pressure head of emitter of any lateral shall be calculated based on discharge requirement of each emitter (from standard table).
- The head loss in the lateral length between the first and last emitter shall be within 10% of the head available at the first emitter.
- The friction head loss in main and sub-mains shall not exceed lm/100m of length. The friction head loss in main and sub-mains can be calculated by Hazen-Williams formula given below:

hr = 10.68 x $(Q/C)^{1823}$ x $D^{-4.87}$ x (L+Le)

Where hr = Friction head loss in pipe (m)

Q = Discharge (m^3/sec)

C = Hazen-Williams constant (140 for PVC)

D = Inner diameter of pipe (m)

L = Length of Pipe (m)

Le = Frictional loss for accessories and fittings, estimated in equivalent of straight pipe (m).

3. Selection of machineries like filters, valves, injector etc:

Depending upon the requirement of greenhouse technology necessary machinery can be incorporated in the proposed drip irrigation system. The head loss (as furnished by the manufacturers) due to these machinery should be added while estimating the total head loss.

Step - 4

1. Estimation of total water required by the system:

Total water requirement of greenhouse/s including open field (if any)

= WR x number of plant

2. Estimation of Horse Power of pump set:

The Horse Power of pump set required for a drip irrigation system is calculated on the basis of estimated peak discharge and total operating head. The total operating head is the sum of total static head and frictional head loss of the system, which includes depth of water, drawdown, frictional loss in pipes, bends, valves etc. and minimum head required over emitters.

So the HP of a pump = (Q x H) / (75 x e)

Where- Q = Discharge (lps)

H = Operating head (m)

And e = Pumping efficiency (0.6)

7.6. Fertilizer Application and Fertigation In Greenhouse

It is mandatory to maximize the benefits of fertilization in the form of plant growth or yield in a greenhouse situation. Proper and optimum application of plant nutrients in natural and/or chemical forms most acceptable to plants is the basic job of fertilization.

In general, 90% of dry weight of a plant consists of Carbon, Hydrogen and Oxygen and the rest 10% includes essential elements like nitrogen (N), Phosphorus (P) and Potash (K) (Table - 28).

Table 28 : Presence of essential nutrients in plant on dry weight basis.

Nutrient Element	ChemicalSymbol	Classification	% of dry weight
Carbon	C	Non-fertilizer	-
Hydrogen	H	Non-fertilizer	89.0
Oxygen	O	Non-fertilizer	-
Nitrogen	N	Primary macro-nutrient	4.0
Phosphorus	P	Primary macro-nutrient	0.5
Potassium	K	Primary macro-nutrient	4.0
Calcium	Ca	Secondary macro-nutrient	1.0
Magnesium	Mg	Secondary macro-nutrient	0.5
Sulfur	S	Secondary macro-nutrient	0.5
Iron	Fe	Micro-nutrient	0.02
Manganese	Mn	Micro-nutrient	0.02
Zink	Zn	Micro-nutrient	0.003
Copper	Cu	Micro-nutrient	0.001
Boron	B	Micro-nutrient	0.006
Molybdenum	Mo	Micro-nutrient	0.0002
Sodium	Na	Micro-nutrient	0.03
Chloride	Cl	Micro-nutrient	0.1

Unlike the plants in an open field, as these plants grow in restrictive environment the volume and type of the growing medium is not sufficient to buffer the short supply of plant nutrients. If the growing medium is soil less, the importance of fertilization, including both macro and micronutrients, increases manifold. Thus, in comparison to open field, the greenhouse crop requires high precision fertilization mostly through drip irrigation system. Organic matter is not generally used in such cases.

However, in case of medium technology or medium cost to low cost greenhouse, where fertilizer application through irrigation system is not so improved and efficient, soil application of organic matter, micronutrients and sometime phosphorus may be adopted. The nitrogen and potassium may be applied through irrigation water.

7.6.1. Plant Nutrient Management and Fertilizer application Schedule

As said earlier, organic matter @ 10-15 kg/m^2 in the form of compost or FYM duly treated with fumigant, may be applied in soil at the time of preparation of bed. However, in most high technology greenhouses where practice of soil pasteurization or fumigation is done regularly the need for preparation of soil with untreated organic matter is eliminated. Instead, all sorts of secondary macronutrients and micronutrients are supplied through irrigation system

However, in case of soil-based medium, along with organic matter, the correct pH level of the medium has to be maintained, which generally turns acidic in greenhouse condition. Finely ground limestone is applied to the medium to raise the pH level. Dolomite limestone is best for this purpose because of presence of magnesium as essential nutrient along with calcium. The rate of application varies with type of root medium and magnitude of the pH correction needed. In general 2.5 to 11.5 kg of lime/m^3 soil medium may be applied depending upon the requirement.

Where pH of root medium is neutral, instead of limestone 2.5 kg of Gypsum (Calcium sulfate) should be incorporated into each cubic meter of root medium every year. This is sufficient to supply entire calcium and almost entire sulfur to the plant throughout the year. If the magnesium status of the crop (perhaps through foliar analysis) is low, occasional application of this nutrient through 250gm of Epsom salt (magnesium sulfate) dissolved in sufficient water is done to treat one m^3 of root medium.

Next is the status of phosphorus in the soil, which is determined through soil test. Unless it is high in the root media, application of 1.5 kg of single super phosphate per m^3 of root medium is needed. Single application will provide necessary phosphorous to the soil medium for a year.

Micronutrient mix, containing iron, manganese, zinc, copper, boron and molybdenum can be applied directly to the soil medium once in a year. Otherwise liquid application of commercial mix of micronutrients can be made through drip irrigation system 3 to 4 times during a year.

The rest two important macronutrients viz. nitrogen and potash may be applied through drip irrigation system in the greenhouse.

7.6.2. Fertigation

It is the application of fertilizers in soluble form through irrigation system. The standard practice is to dissolve high analysis fertilizer carriers into concentrated solution, mixed up in necessary proportion. This is called fertilizer 'stock'. This 'stock' is injected in specific proportion into the water line of the irrigation system of the greenhouse to obtain final concentration required for application in crop.

Generally liquid fertilizers containing specific plant nutrients are applied to plants of greenhouse through drip irrigation system in two different ways.

1. **Constant feed**: This application entails administering low concentration of fertilizer each time the plant is irrigated. It is the most popular method of greenhouse fertilization. Through this method plants receive a fairly constant supply of nutrients in the root medium for sustained growth and development.

2. **Intermittent feed**: Greenhouse plants may be fertilized according to a periodic schedule such as weekly, fortnightly or monthly. The disadvantage of this method is the level of nutrient available at the time of application, which is higher than that of plant requirements. This gradually decreases over time, and touches a lowest level just before the next application. Naturally the plant growth is not continuous but fluctuating (Fig.-54).

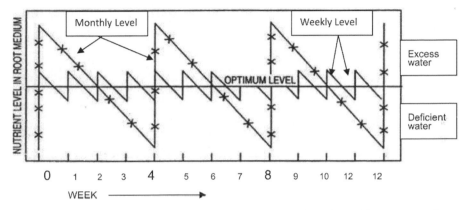

Fig. 54: Nutrient level status of root media during weekly and monthly programs of fertilizer application

7.6.2.1. Forms of Fertilizers for Fertigation

Fertigation essentially requires water-soluble solid or liquid fertilizers. The fertilizers, which are applied directly into soil, are commonly not completely soluble in water and also contain some non-soluble solids, as filler, which makes them unsuitable for application through drip irrigation system. However, two types of fertilizers can be used for fertigation purpose.

1. **Liquid fertilizer:** It is the solution of one or more plant nutrients in liquid or suspension form. It generally contains basic plant food elements suited to crop need and 'tailor made' to farm requirements. Due to lack of infrastructure facility to produce and handle such fertilizers and lack of application facility in the field these types of fertilizers have not been popular in India.

2. **Water-soluble fertilizers:** The water soluble fertilizers in solid form carrying two or more major nutrients like N,P,K and sometime micronutrients are generally used for fertigation. For fertigation purpose urea, ammonium nitrate (NH_4NO_3), potassium nitrate (KNO,), and phosphoric acid are used as source of Nitrogen, Potash and Phosphorus respectively. Solubility and purity of solid fertilizers, used for fertigation purpose is the most important aspect.

Several factors may inhibit proper mixing and impose problem due to incompatibility among fertilizers, including water. The possible remedies are narrated below:

1. First the **procedure** should be standardised. The required solid fertilizer is added with a small quantity of water and mixed thoroughly. This concentrated solution is then sieved and transferred to a mixing container, which is filled up with 50 to 75% of the required water. After that the rest amount of water is added to fill up the container. This final mix is known as 'stock' and is added to the irrigation water.

2. An acid or acidic fertilizer shall be never mixed with chlorine compounds like hypochlorite or with compounds containing sulphate or calcium.

3. Checking with chemical supplier for information about insolubility and incompatibility.

4. Checking the water quality and never to use hard water. Water having pH 5.5 to 6.5, EC less than 0.1, low level of carbonate, bicarbonate, sodium, chlorine and free from heavy metals is considered good for irrigation.

7.6.3. Methods of fertilizer application and fertigation

There are a number of techniques to add fertilizers into the irrigation water. The irrigation systems may be (1) hand watering with containers (2) sprinkler, (3) Micro-sprinklers and misting and (4) drip irrigation through which fertilizers can be applied to crop.

While fertigating through any of the above systems, it should be ascertained that the optimum concentration is maintained taking into consideration the crop reaction. The 100% water-soluble solid or liquid or both type of fertilizers, as per requirement, are to be added in a bucket full of water. This high concentration fertilizer solution is then mixed with measured water in a mixing tank and the mixture is called as 'stock'. This mixture or 'stock' is then required to be added to the irrigation water in a specific proportion five minutes prior to use.

Fertigation is commonly and regularly done through drip irrigation system with the help of some additional equipment like pump, valves, ventury or infector etc. The size and the capacity of the additional equipments depend upon the concentration grade and frequency of fertilizer application. Normally, less amount of fertilizer solution with more frequent application require smaller and less costly fertigation units. The different equipments used for fertigation are described below:

1. **Ventury**: It is a low cost fertilizer injection device used for the irrigation system. A proportion of irrigation water is diverted from the main water line and passed through a device, which increases the velocity of the water flow. The water enters the siphon-mixing unit at a very high velocity, generating a suction pressure (Siphon) in a feeder line that is dipped into the soluble fertilizer concentrate. The suction draws the fertilizer into the watering hose, where it mixes with irrigation water according to a pre-determined concentration (Fig. 55).

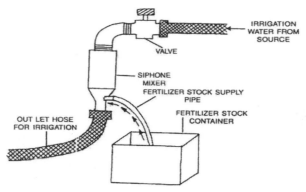

Fig. 55: Ventury - a low cost fertilizer injection system for drip irrigation

2. **Fertilizer tank**: It is almost similar to the ventury system. Here the diverted irrigation water passes through a tank containing liquid or soluble solid fertilizers then returns to the main line. A pressure reducing valve is placed in the return pipes causing dilution of fertilizer in the tank and flow of diluted fertilizers into the main irrigation line.

3. **Fertilizer injector**: This is a piston or diaphragm pump, which is driven by the water pressure of irrigation system, and as such the rate of injection is proportional to the flow of irrigation water. Through this system a high degree of control over the fertilizer injection rate is possible.

7.7. Plant Protection Measures

The greenhouse environment with higher humidity and temperature is prone to infestation of animal pests and pathogens. The transportation of pest and causal organism of disease into the greenhouse occurs in different ways. This may be via plant material, intruder's clothing, containers and other materials. Besides, there are wind, water and soil borne pests and diseases. The wind borne pests (as vectors) and spores of disease causing organisms generally intrude into the greenhouse through leakage or openings of walls and roof. In microenvironment of greenhouse it is very difficult to protect the crops from infestation of pest and diseases. So a careful plan is needed for prevention and control of insects and pathogens, which attack or damage the crops.

7.7.1. Prevention for control

Prevention plays a very important role, particularly to control the disease infestation. Several preventive measures are used for greenhouse cultivation to control infestation of pests and diseases. Some of them are briefly discussed below:

(i) **Obstruction of entry of insect:** Covering all the openings of greenhouse by insect proof (IP)-net is the best method of minimizing the entry of insects, particularly the smaller insects. It can reduce the pest and disease infestation up to zero level.

Some of the IP-nets (70 mesh) can prevent insects as small as thrips. Prevention with application of 40 mesh IP-net for whitefly or such insects, which act as vectors and carry disease-causing viruses, reduce the infestation of such diseases.

The door of greenhouse should be designed in such a way that the insects or spores do not get any free access inside. Generally a pre-entry chamber is made in front of the door to avoid direct entry of people or equipments.

(ii) **Checking of Plant Material:** At the time of purchase and before planting, thorough **inspection of plant** material should be done to prevent entry of pathogens and insect (in the form of egg or larva). It is better not to purchase plant material produced in nursery infested with pest or disease or seeds from un-reputed organization.

(iii) **Weed Control**: Many weeds harbor insects and pathogens that ultimately enter the greenhouse and create serious problem. It is important to **remove weeds** in and around greenhouse, either manually or by application of selected herbicides.

(iv) **Avoiding foreign soil or non-pasteurized soil or root media**: Soil borne pathogens and eggs or larva of insects are present in almost all types of soil, which can cause serious damage to the crops of a greenhouse. This soil can be transported though bottom of the shoe, dirty implements etc. Proper scrapping and washing of outside soils attached with the things that enters into the greenhouse may avoid this problem.

(v) **Cleaning up of debris**: The debris, particularly the pruned or clipped or detached plant parts should be disposed off immediately. The juicy green plant parts can act as a very good medium for growth of pathogens.

(vi) **Sterilization of containers and other tools**: Sterilization of containers and other tools reduces spread of diseases from open field or from one greenhouse to another greenhouse. It is a must when an epidemic breaks in the area.

(vii) **Pasteurization or fumigation of root-media**: It is narrated in detail earlier in this chapter (6.3.5) in preparation of root-media.

(viii) **Early detection**: Attack of pest and diseases still occurs in the greenhouses after taking all initial preventive measures. Detection at an early stage helps to prevent any significant spread of pests or diseases. The early detection of disease or pest infestation can be noticed by identifying early symptoms of diseases, egg laying or colony structure of the pests, nature of initial damage by the insects and pathogens, critical number of pest count and presence of excreta of pest in greenhouse.

Naked eye or magnifying glass can be used for the above mentioned detection purposes. For early detection of insects, in physical form, 'traps' should have been installed inside the greenhouse. Microscopic detection may be done to detect the pathogen of different kind.

(ix) **Physical verification schedule**: In every occasion of training, pruning of the crop and other inter culture operations the plants shall be inspected properly to detect any kind of infestation of pest or disease.

(x) **Application of Chemicals and bio-pesticide**: Depending upon the outcome of the earlier detection procedures, if necessary, chemicals or bio-chemical materials may be applied to stop the spreading of pest and diseases in a greenhouse by way of killing the causal organisms or pathogens. However, residual toxicity of different chemicals and growing of resistant power of different pests against specific chemicals are the two problems to be addressed properly.

To avoid residual toxicity in crops, choice of pesticide and corresponding danger/ risk-period, plays a crucial role. Chemicals of less residual toxicity targeting specific pest should be preferred. Bio-pesticide is always preferred over chemical pesticide, due to the presence of other chemicals along with the poisonous bio-molecule and its degenerating capacity. If required these toxic chemicals should be chosen considering the harvesting-interval of crop having periodical harvesting schedule like most vegetables and waiting-period of pesticide.

However, after having made proper identification of insect pest or pathogen and understanding their life cycle and biology, the grower must have a position to take wiser decision in choosing the best category of chemicals to control the pest or pathogen.

In case of some specific diseases/pathogen, the problem can be solved by adopting proper prevention methods like seed treatment, particularly with systemic fungicides or bio-fungicides. Otherwise, for a given disease problem of greenhouse the fungicides may be applied at short intervals to unaffected plants to prevent the pathogen from entering and spreading.

The pesticides, particularly the bio-pesticides are generally applied in the late afternoon or evening when temperature is descending. The risk of phyto-toxicity is greater when applications are made during the middle of the day due to high temperature. Insecticides should not be applied when the plants are under water stress.

Since recent past pesticides are specially formulated for greenhouse application, to protect specific crop against specific or specific group of pests.

7.7.2. Strategies of control of greenhouse pest in naturally ventilated greenhouse

To be effective, the timing of application of the pesticide is critical. The following four factors are important to decide the pesticide application in a greenhouse; (1) the residual life of pesticide, (2) the life cycle of the insect-pest or pathogen, (3) mode of action of pesticide/fungicide/miticide/bactericide/etc, (4) method of application.

Proper combination of the above-mentioned factors can effectively control any specific pest or disease. However, strategically, first one has to identify or select a single or a group of pests/diseases against which the control measure shall be taken. Then the life cycle of that insect or pathogen and its stage of growth or development, at which it is most susceptible to control, should be properly identified. At the same time the method of application (e.g. smoke, spray, soil application etc.) as well as the type of chemicals to be used should also be decided carefully.

Considering all the above aspects the scheduling of application will be finalized. However, it is evident that intervals of five to seven days are generally effective in controlling many greenhouse insect and mite attacks.

Special reference of Controlling Mites: Mite populations are held in balance by natural enemies, weather and host quality. Mites are known to develop resistance to pesticides (miticide or acaricide) rapidly. They also exist as a heterogeneous mixture of growing stages in a given population. To control mite effectively three-miticide (of different mode of action) cycle is recommended.

7.7.3. Specific strategies of Controlling of disease causing pathogens of greenhouse

In an artificially controlled cultivation environment, disease can be prevented by several cultural practices; (1) environment control; (2) strict observance of sanitation; (3) use of healthy disease free plant materials and (4) use of sterilized soil and other soil additives.

Solar Trap

Light Trap in greenhouse

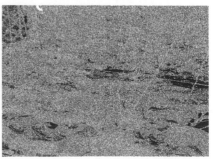

Sticky trap in Greenhouse   Marigold for detection of pest

As the greenhouse condition is sometimes very much favourable for infestation of some pathogens due to high status of temperature and humidity, chemical control is often necessary to restrict spreading of disease throughout the greenhouse.

7.7.4. Methods of application of pesticide in greenhouse

Not like open field situation, there are a few specific methods of direct pest control that are suitable for greenhouse crops.

1. **Spraying**: Most greenhouse pests are controlled by spraying appropriate chemicals in the form of emulsifiable concentrates (EC) or wettable powders (W.P.). Certain pesticide formulations are approved for use in greenhouse.

2. **Aerosol**: Aerosol is a readymade sprayer/atomizer, containing necessary pesticide, usually applied in the greenhouse when immediate killing of pests of a specific crop is necessary. With this spray, very little residue is left on the plant and only the upper surface of plant shows this residue. Aerosol may be applied to dry leaves on a calm condition minimizing the air movement inside. After application, the greenhouse usually kept closed overnight.

3. **Fog**: As dusting is not commonly practiced in greenhouse, pesticides are applied in the form of fog using fogging equipments. Fogs are oil based and usually prepared at I0 % strength of the regular insecticide or fungicide. The fogging equipment heats up the pesticide, which breaks down into a white fog like gas that spreads throughout the greenhouse.

4. **Smoke**: Unlike sprays and fogs, smoke does not require any special equipment for application. Instead a combustible formulation of pesticide packed in containers is placed in the center of the greenhouse and ignited. Smokes are generally not as phytotoxic to foliage as other gas applications. At the time of application the greenhouse must be closed completely.

5. **Soil application**: Soil borne disease and pests may be controlled by drenching with pesticide, granules or powder. Formulations may be applied to the soil surface and washed down by irrigation water.

Since recent past few pesticides are specially formulated for greenhouse application, by the developed countries, to protect specific crop against specific or specific group of pests.

A common list is given mentioning few insects and pathogens of greenhouse and their respective controlling chemicals.

Table 29: Different pest control materials (chemicals, plant extract, bio-control, etc)

Name of common pest	Chemicals, Plant extract and Biological Control
Insect pest	
1. Aphids.	Melathion, Diazinon. Endosulfan, Plant (neem + karani) extract, and lady-bird beetle, wasps citic as bio-control.
2. Mites.	Dicofol, Kelthen, Tetrasan, Sanmite, Sultan, Akari, in suitable rotation.
3. Mealybugs.	Malathon, Diazinon or Orthene with a surfactant.
4. Leaf miners.	Fenthion/Triazophos + Decamethrim.
5. Thrips.	Dimehoate, Imidacloprid,- Fipronil etc.
6. Whitefly.	Malathion, Rotenone or Orthene etc.
7. Nematodes (root-knot)	Halogeneted Hydrocarbons, Carbofuran, neem/mahu cake.
Disease pest	
1. Bortrytis Blight	Captan, Mancozeb, Zineb etc.
2. Powdery mildew	Sulfar, Benomyl, carbendazim, Fenarimol etc.
3. Root rot.	Captafol, captan, Thirum etc.
4. Damping off.	Captan, Thiram, etc. (for seed treatment)
5. Verticilium wilt.	Streptocycline (for seed treatment and early stage spraying)

7.7.5. Types of major disease and insects with their control methods

7.7.5.1. Types of disease pest of greenhouse

There are numerous pathogenic diseases of greenhouse crops. However, in greenhouse condition some specific pathogens play active role to cause and spread such diseases and damage the crop. In general the pathogens are categorized in four different groups, viz. virus, bacteria, fungi and nematode.

(i) Viruses: Viruses are minute microscopic organisms similar in size and chemistry to the genetic material (mostly, DNA) present within the plant and animal cell. It lives within the cells and causes abnormalities to the host. No chemical or process is available to destroy viruses without damaging the host. So the only procedure to eliminate virus is to remove and destroy the infected plant. Few prominent viral diseases are leaf curl virus, mosaic virus etc.

Control: Spreading of virus can be prevented. In general insects spread viruses while they are feeding on a healthy plant after feeding an infected plant. Prevention of these insects into the greenhouse reduces the scope of infestation by virus. Grafted plants can be infected by virus through rootstock or sion. Use of resistant variety is a useful method to avoid virus infestation.

(ii) Bacteria: It is a single cell microorganism. Disease causing bacteria causes severe damage to the crops. After infestation it is very difficult to control. Bacterial disease spreads very fast. Some common bacterial diseases are bacterial wilt

(*Pseudomonas* spp.*)*, bacterial blight (*Xanthomonas* spp.)*,* bacterial soft rot (*Erwinia* spp.), crown gall *(Agrobacterium* spp.*),* etc.

Control: Control of bacterial infestation is generally done through prevention (as described earlier) and elimination of infected plants. A few bactericides are available which are generally used through some prevention purposes like seed treatment, seedling dipping etc. Application of bactericides like *Streptomycin Sulfate, Streptocycline* (@0.25%) in early infestation may prevent spreading of the bacteria in healthy plants.

(iii) Fungi: It is the most common type of pathogen that frequently attack crops both in open field and in greenhouse. Fungal organisms are physiologically much more complex than bacteria. They are multi-cellular organisms often comprising of several tissues. The numbers of fungal diseases are more than any other types but they are also susceptible to best control measures.

Control: There are a range of chemicals known as fungicides that can effectively control the infestation of disease. Some of them are pathogen group specific whereas some others are broad spectrum i.e. effective against a range of pathogens.

(iv) Nematode: Nematodes are very small, rather microscopic, round worms, sometimes called as eel-worms. It is present in almost all soils but most of them are not harmful. When large population of harmful types builds up in roots, it affects the growth of crops by preventing normal root functions. Root-knot nematodes (*Meloidogyne*) is the most common disease that produces symptoms like stunted and unthrifty growth with a tendency to wilt on warmer days. The root galls of affected plants are conspicuous and easily recognizable. Nematodes are transported from one place to other through soil or infected plant parts. Migration of larvae through soil is limited to a few feet per year.

Control: Soil pasteurization or solarization is the most effective method of eradicating nematodes. The root attacking nematodes can also be controlled effectively by application of suitable organic substances like neem or mahua cake @500kg/ha before planting and non-fumigant chemicals like Carbofuran @ 1 to 2 kg a.i. /ha at the time of planting.

Some of the other diseases with symptoms and control are stated below

1. Powdery mildew: It is characterized by the presence of whitish powdery growth on the surface of leaf, stem, inflorescence etc. Heavy colonization of this pathogen in a small area of a leaf kills the cells developing black-patches. The mildew spores are easily detached and are carried by air currents to surrounding plants.

Control: Several fungicides like Copper Sulphate, etc can control this disease but Sulfur is the most common and effective. Bacteria like *Bacillus pumilas* and *B. subtillus* are effective to control mildew diseases in greenhouse.

2. Blight: *Botrytis* spp. is the causal fungus of this disease and probably causes more damage than any other single pathogen. Greenhouse condition is best suited for this pathogen, if water is available on plant. It causes brown rotting of plant tissues blighting the affected area. Leaves and stems are commonly affected. Spores are dislodged and carried out by wind to healthy plants. However, without water the spores cannot germinate to produce new infestation. *Control:* No splashing of water on plant surface is the first control measure. Increasing air circulation and reducing high humidity is the second line of defense against blight. Many fungicides can effectively control its infestation. However, at the time of selection of fungicide it should be noted that the chemical does not hamper the growth of the plant. 'Captan' spray is good in all these aspects. However, Scala(*Pyrimethanil*), *Streptomyces Griseoviridis,* etc are the best for greenhouse application.

3. Root-rot and damping-off: *Rhizoctonia* and *Pythium* spp. together with *Thielaviopsis* spp. are mostly responsible for root rot diseases. Basal stem rot of older plants, damping off of young seedling are the common types of this disease. These pathogens are common inhabitants of soil. The *Pythium spp.* and *Rhizoctonia spp.* grows in cool wet and hot moist condition producing black and brown colour on affected parts respectively. *Rhizoctonia* is mostly responsible for seedling damping of

Control: Drenching of soil with formaldehyde before sowing or selected fungicide or combination of fungicides at the time of sowing give effective control against the infestation of such diseases. Seed treatment with *Captan or Thirum @ 2 grams/kg* of seed should be done before sowing. However, spraying of *Mancozeb,* Tetrraclor(*PCNB*), *Tricoderma Virens/Harzianum,* etc are best for greenhouse application.

7.7.5.2. Types of insect pest of greenhouse

Several types of insect pests attack plants in the greenhouse. Major infestations are due to some specific types of pests. Basically these pests feed on the plant and complete life cycle with the help of the plant. Thus, the feeding habit and life cycle of these pests should be known properly. The controlling measure is very much crop-pest relationship specific and are also diverse in nature. However, insecticides with systematic action are limited for greenhouse use except for seed treatment in very early stage of growth. Some important pests are detailed below.

(i) Aphids: There are several types of aphids differentiated by colour. Green peach aphid (*Myzuspersical*) or Sulzer is the most prevalent greenhouse aphid, particularly the wingless forms. The colour of winged forms is brown whereas the wingless forms are yellowish green or pink to red in colour. They feed by piercing mouthparts and suck out the cell-sap.

Control: For biological control lady-bird beetles, wasps, lacewings and other such insects may be released into the greenhouse. However spraying of Malathion (0.1%) or Endosulfan (0.05%) may be done to control the aphid infestation.

(ii) Leaf miners *(Liriomyza trifolii)*: The larval stage of this insect makes unsightly tunnels between the outer layers of leaf. The adult female is a stocky fly, which punctures the leaf surface with a tube like appendage on the abdomen known as 'ovipositor' and inserts eggs through it. The larva called maggot (1/10 inch) do such tunnels on feeding. About 5 weeks is required to complete the life cycle from egg to adult out of which two weeks are used for tunneling.

Control: Release of natural enemies like *Diglyphus spp.* may control leaf miner effectively. In severe infestation, spraying of Fenthion (0.01%) or Triazophos (0.04%) with Decamethrin (0.0014%) may be done.

(iii) Mealy bug: Mealy bugs (Pseudococcus) are oval-shaped that appear white because of the wax like powder, which covers their body. They feed by means of a piercing, sucking mouth part.

Control: Many natural enemies feed on and kill mealy bugs in open condition. The waxy protective layer makes it difficult to control them by any pesticide, thus use of surfactant increase the efficacy of the pesticide. For greenhouses, where it is not physically possible to remove mealy bugs and where biological control may not be feasible, spot treatment with Insecticidal soaps, horticultural-oil, or neem-oil insecticides may be used to suppress the infestation. Insecticides like *Dinotefuran, or Pyrethroids*, may control mealy bug but not much effective than soap and oil spray and can be devastating to beneficial pests, thus, better to avoid such insecticides.

(iv) Mites: These are not insect and belong to class '***Arachnida***'. There are many species of mites, including several that attack crops. Many are very small, about 1/100 inch of size *(Stencotarsonemus pallidus),* and cannot be seen with unaided eyes. High humidity (80% or more) and low temperature is favourable to them. The symptoms of mite attack are curling of leaflets from outside inward and distortion of young leaves. Red spiders (*Telranychus urticcie*) are a kind of mite that perhaps is the most troublesome of all greenhouse pests. They may be greenish or yellowish or reddish in colour and have two dark spots on their bodies. These two-spotted mites place themselves under the leaf and cause chlorotic stippling of leaves. The spiders spin a silk strand, which forms a web over leaves and flowers. The leaves and flowers soon begin to dessicate and consequently turn brown.

Control: Use of IP-net in openings can prevent the entry of mite into the greenhouse. Spray type irrigation system is detrimental to mites. Suitable pesticides generally termed as 'miticide' may be sprayed, two or three in a cyclic schedule, to control the mites on a continuous basis. In high temperature it may be done in an interval of every two days. Spraying 'Dicofol' @0.04 % can control the two spotted spider mites. As mites easily show resistance against repeated application of any particular chemical it is always better to use three types of miticides in rotation throughout the lifecycle of a particular crop. For the mites at immature-stage the miticide control-cycle is Tetrasan (etoxazole)— Hexygon (Hexythiazox-50%) — Judo (Spiromesifen-45.2%) and in mobile-stage the cycle may be Avide (abamectin)—Sanmite (Cyflumetofen-18.7%) /Pylon (Chlorfenapyr-21.4%). Apart from these, several other new and very effective greenhouse miticides are available in the market like Sultan(Cyflumetofen), Akari 5SC(fenproximate-5%), etc. Surfactants are used with most of the miticides.

Otherwise proper schedule of suitable bio-pesticide/miticide or combination of bio-pesticides, available in the market, may be adopted to control greenhouse mites. Most of these are neem/karanj oil/extract based mixture.

(v) Thrips *(Scirtothrips dorsalis*): Thrips are small (1/25 inch) insects with two pairs of fringed wings which move by air current. It feeds on plant parts, commonly buds, petals, axils of leaves etc, of a broad range of crops. They can be detected by tapping buds or flowers over a sheet of white paper. Now-a-days thrips is a severe problem in greenhouse cultivation.

Control: Application of IP-net of higher mesh in all the openings of the greenhouse can prevent thrips from entering. Higher humidity and regular application of water through misting may reduce the infestation. Very selective pesticide (Dimethoate, Imidacloprid, Fipronil etc.) can control thrips effectively. As biological control a predacious mite *(Neoseiulits cucumeris)* can be effective but require several weeks.

(vi) Whitefly *(Bemisiu tubaci)* : It is a wide spreading small insect that carry sprays viral diseases and is very tough to control.

Control: Covering the openings of a greenhouse with IP-net is an effective measure to control white fly. Apart from common prevention measures, use of parasitic wasp *(Encarsia Formosa)* or bio-agents like *Verticillium lecanii* is quite effective. Insecticidal soaps, neem oil, or petroleum-based oils control only those whiteflies that are directly sprayed on. The soil-applied systemic insecticide *imidacloprid* can control whitefly nymphs but damage the beneficial pests. However during active period spraying of *Malthion* (0.01 %) or *Imidacloprid* or any new generation greenhouse pesticide may be tried as one shot. Otherwise, on regular basis, soap and neem-oil based control is the best way.

(vii) Other greenhouse insects: Scale insects and worms are the other pests, which attack greenhouse crops frequently.

7.8. Canopy Management and other Crop Husbandry Practices

Now-a-days among different precision farming technologies most important one is crop-canopy management. The other precision technologies regarding water, nutrient, plant protection, etc are almost available on a standard package but, it is really difficult to have a standard canopy management programme for a crop. It needs enough experience to take on-spot decision to do it. It involves pruning, training, staking, trellising, etc. Furthermore, operation like pruning involves reduction of shoot, root, leaf, fruit, etc of a crop in different stages of its growth and in different magnitude.

7.8.1. Principle of pruning and training

Nowadays the modern crop, particularly the horticultural crop management requires most unique methods of pruning and training. For greenhouse cultivation it is almost mandatory to accommodate more number of plants per unit area of greenhouse mainly through canopy management. These manipulate the natural growth of crop and accordingly provide more yield. Proper pruning and training require more individual capacity and experience and is not possible by any thumb rule procedure. However, the following principles may be applied to get the maximum benefit from pruning and training.

- **Exposure of maximum leaf area to sunlight**: This creates more scope to increase photosynthesis that ultimately enhances growth and yield. This also allows relatively less intensity sunshine to penetrate inside the canopy for more photosynthesis.

- **Vertical development of crop**: Vertical development of crop canopy through pruning and training is a key factor to increase the yield of a specific crop. This modified canopy can be accommodated in smaller areas by way of increasing the exposed leaf-area : ground-area ratio. If a plant is supplied with more than sufficient growth inputs it cannot utilize the same unless its leaf-surface-area is increased to entrap more solar energy. This excess entrapping of solar energy by crop canopy may be termed as "corrugated effect".

- **Discard selected biomass and increase the yield to biomass ratio**: Plants generally maintains excess biomass for many purposes other than growth. Frequently a good percentage of leaves present beneath the upper layer of the canopy cannot open stomata properly for optimum photosynthesis (Fig. 56).

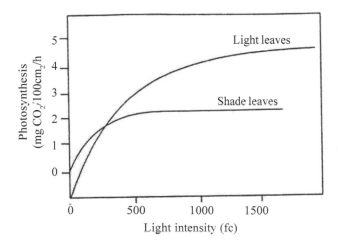

Fig. 56. Different level of photosynthesis between leaves present in light and shade.

This excess biomass acts as defense mechanisms for the plant. This plant parts/ leaves have an important role against the risk factors the plant may face during its growing period. These risk factors may be shortage of water and/or nutrients, breakage or destroying of branches or any plant parts due to pest, storms etc.

Now, as the greenhouse technology itself takes care of all the above-mentioned risk factors, these excess biomasses practically have no purpose. Instead, the food and energy required to maintain this excess biomass, can be diverted for more production.

- **Enhancement in emergence of healthy sprout/bud/fruit**: In off-season, perennial plants store food (photosynthates) in its stem, and that is utilized for its early growth in the next season. In case of annual plants, frequently, it is found that flower/fruit close to the main stem yields better, both in quantity and quality, than lateral branches (tomato, pepper *etc*). So by proper pruning of the stem or branch the quantity and quality of yield can be increased. With the help of this method, growing season of the crop can be changed according to the demand of market.

7.8.1.1. Training

It is the practice performed on a plant that gives a specific retainable structural frame to that plant by way of regular pruning. Actually the direction and spacing of selected branches or main stem of a plant is managed through pruning or otherwise in such a way that they provide a proper frame allowing more sunshine and air movement. This increases the efficiency of biomass and reduces the land area covered by canopy of that plant (to accommodate high density planting).

7.8.1.2. Pruning

Pruning generally refers to removal of plant parts, especially shoots, branches, buds, roots or leaves.

Up to recent past the term 'pruning' was associated with trees or orchards and in minimum scale it involved removal of dead, diseased, broken and dried plant parts. However, proper and judicial pruning gives a proper shape, allows more trapping of sunlight and encourages the development of secondary branches, thus increasing the yield. But this is not all, nowadays pruning is done more methodically and scientifically to improve quantity and quality of yield significantly. In some ornamental plants like Rose etc pruning is used to make the shape more manageable and also to get quality flowers.

7.8.1.3. Use of Training & Pruning for greenhouse cultivation

Training & Pruning of perennial crops

1. In off-season, perennial plants store food (photosynthets) in its stem, and that is utilized for its early growth in the next season. So judicial pruning of stem encourages more healthy and productive shoots.

2. Decrease in shoot growth and cambial activity in off-season makes more food available for its growth in next production season.

3. Proper pruning of branches during dormant season causes more fruit to set on the branches left.

4. Pruning should be in proportion to the root system.

5. Removal of 30 cm or more terminal stock at the time of harvesting or plucking is helpful for growth of mature buds for quality production or more new twigs for more yields and indirectly served the purpose of pruning.

6. Root pruning is done after drying out the soil, which induces flowering and fruiting and determine the time of flowering.

7. De-budding or pinching off initial vegetative growth frequently produce quality flower and fruit.

Training & Pruning of annual or bi-annual crops

1. Stretch the frame and canopy vertically to increase the leaf area exposed to sun to trap more sunlight.

2. More leaf surface per fruit/flower may not be the only way to improve the yield. Proper pruning of excess biomass always increases the size of fruit, especially in plants that set of excess or heavy fruit/flower.

3. Pruning may help to conserve water by reducing foliage that loses water through transpiration.

4. The un-pruned plants spread over larger ground area occupy more greenhouse area.

5. Reduce the excess biomass that does not take part directly in crop production.

6. Pruning of the initial bud/ flower may induce healthy and productive lateral branch.

7. Judicial pruning or pinching of the lateral shoots force the plant to grow vertically and produce better crop in the main shoots.

8. Proper pruning and training can accommodate much higher number of plants thus increasing the yield.

9. Fruit pruning (thinning) can overcome the negative effect of over-bearing or excess fruit-load and also maintain the fruit quality.

7.8.2. Staking and Trellising

It is an age old practice performed on different climbers and creepers in open field situation. It was done to restrict its growth to our reach and to give them more space to grow and yield more (Fig. 57). Almost the same principals with very high precision, is carried out for greenhouse crops.

After execution of proper training, many shrub type crops require support to develop vertically. Depending upon the crop either staking or trellising is necessary to push or pull the canopy upwards. In case of climbers or creeper it is easier to stretch the crop canopy upwards, but in case of freestanding shrubs the upward stretching is difficult and require more technicality.

For freestanding bushy crops trellising is the best method where the vertical wire or rope is used to pull the crop canopy upwards judicially. Finally this crop canopy with fruit load hangs from these wires or ropes. Type of crop and its physiological character is a vital aspect to design the trellising method.

Fig. 57. Different types of stacking and trellising system practiced in open field

7.8.3. Specific training & pruning method of some greenhouse crops

Tomato: Greenhouse grown tomato cultivars are commonly of indeterminate type, thus they need regular training and pruning within a few days after transplanting. Each of these plants are supported by plastic or non-binder twine, loosely anchored on the base of plant with the help of non-slip loop or plastic clip. Removing all the side shoots or suckers should retain only the single main stem (Fig. 58.a). This can be done by means of snapping, particularly in the early stage. As the plant grows, the twine should be wrapped clockwise around the stem with one complete swirl every three leaves. The twine under the leaves, not under the flower truss or fruit clusters, should support the stem.

When the stem reached overhead supporting wire/structure, untie the twine and release the excess twine at least 2 -3 feet to drop down the stem angularly in one direction for a row and opposite direction of the next row and tie the twine again. For good culture practice, this will require to practice in every 15 to 20 days gap. As fruit mature on the lower parts of the stem, pinch off the older leaves below the fruit cluster. Prior to leaning or lowering of stem about 4 -6 lowermost leaves are to be removed. At the time of pruning/thinning proper care should be taken in such a way that no rough wound or scratch is created in the plant or fruits.

Usually good pollination can set 6 to 8 fruits per cluster. For large fruit cultivars, to obtain quality tomato in respect of size, shape and uniformity 4 to 5 tomatoes per cluster should be maintained by removing the inferior fruits of the cluster. Cluster thinning should be done once a week.

Sweet Pepper: The transplants are initially grown as single stem upto 9-12 leave stage. Then one or two terminal flower develops where the main stem divides into two, and some times more. These terminal bud/s are to be removed just after appearance and the excess of two branch stems are also to be removed. These two branches are allowed to grow as single stem removing all other sub-branches by pinching or cutting, leaving two leaves and one flower at each internodes.

These two stems in each plant are trained individually with the help of strings hanging from overhead supporting wire/structure placed at 8-9 feet height. The stems are clipped with the strings by plastic clip or non-slip loop. The stems are either loosely trellised or bound around the strings. The plants will continue producing terminal flower and two side shoots at every internodes of which one stem shall be pruned leaving one flower and two leaves to continue the stem as single stem (Fig.58.b). If extra leaf area is required for more photosynthesis or for shading the fruit, three leaves may be left in side shoots instead of two. Training and pruning of sweet pepper should be done every three or even two weeks during the period of growth. With proper environment control system this crop can grow upto 8 to 9 feet in 9-10 months period. In the later stage the stems are topped to avoid breaking and to improve fruit size.

Cucumber: The basic principle of training of cucumber plant is to maximize the interpretation of sunlight on leaves. For that matter the plants are trained to grow upward without allowing the horizontal spread of the canopy. Several training systems exist in greenhouse cucumber production depending upon the cultivars, climate, greenhouse-facility, and growers preference.

The main stem of each closely planted plants are allowed to climb along the polythene twine to the overhead wire normally fixed 8 -9 feet above the ground. The twines are anchored loosely to the base of the stem/plant with non-slip noose. The upper part of the twine is alternatively tied to the overhead wire running along the bed/row. As the plant grows it is trellised upward on the twine and after reaching overhead wire the stem is again trained for downward direction. Care should be taken while trellising the stem to avoid any damage to the flower buds.

Through proper pruning no lateral shoots and fruits are allowed on the main stem upto 1,5 to 2 feet above the ground. After that the weak and unproductive laterals should be removed from the main stem. Excessive leaf growth should be discouraged to allow proper colouring of the fruits. The production of fruit depends on continuous development of leaf axils. Multiple occurrences of fruit in a single axil should be thinned to one. 'Fruit thinning' is also necessary to avoid production of malformed and non-marketable small fruits and in case of excessive fruit load. Proper pruning of each plant will be done on the basis of plant vigour and fruit load. To perform this properly a lot of experience is required. For cucumber two basic types of trellising systems are used for greenhouse: Umbrella type and tree-type (Fig. 58 c).

Umbrella System

This system is straight forward, not too demanding in labor and easily understood.

1. Tie the cucumber plant to a vertical wire, 7 feet tall. Pinch out the growing point at the top.
2. Provide support for all fruit that develops on the lower part of the main stem.
3. Remove all laterals in the leaf axis on the main stem.
4. The top two laterals should be trained over the wire to hang down on either side of the main stem. Allow these to grow to two-thirds of the way down the main stem.

5. When the fruits on the first laterals have been harvested, those laterals should be removed back to a strong shoot, allowing the second laterals to take over. Repeat this process for lateral.

6. This renewal system will maintain productivity of plants.

Tree Trellis System

1. Tie the cucumber plants to horizontal wires spaced about 2 feet apart. The top wire should be about 6 feet from the ground.

2. Remove all the leaves and laterals on the bottom 20 inches of the plants.

3. When the main stem has reached the top wire, tie it and remove the growing tip.

4. Allow the laterals at each leaf axis along the main stem to develop two leaves, then cut the growing point.

5. Train the top shoot developing the leaf axis along the wire.

6. When most of the fruit has been harvested on the main stem, allow a lateral to develop as replace-ment and prune in the same manner as the main stem.

Rose: From initial stages only 1 to 2 main stems are allowed to grow to form the basic structural frame by proper training. The frame should not exceed the spacing area allotted to an individual plant. Every year after the production season these main stems are cut back leaving 4 to 5 nodes. Last year's flower producing branches should also be pruned out. All weak stems shall be removed and the wayward stems shall be redirected by proper pruning. For older plants roots may be pruned to promote the development of new young roots. However, in case of high density planting the plants are uprooted, for replacement, every 2 to 3 years. Apart from yearly pruning, the plants regularly get pruned automatically while flowers are harvested with long stalk having 4 to 6 full grown leaves. In general the flowering stalks are cut above the node with last leaf having 5 leaflets. This helps to grow new strong flowering stem from peteoles of the leaves below the harvesting cut (Fig. 58d).

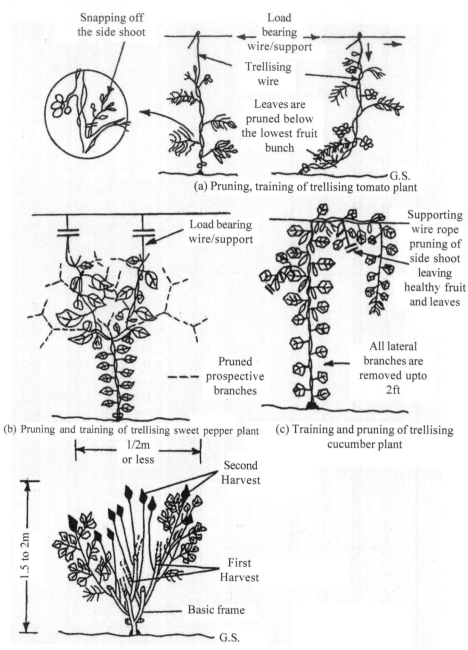

(a) Pruning, training of trellising tomato plant

(b) Pruning and training of trellising sweet pepper plant

(c) Training and pruning of trellising cucumber plant

(d) Rose plant pruned (yearly) to basic frame and subsequent pruning at the time of harvesting.

Fig. 58. Pruning, Training and Trellising of different Crops - (a) Pruning, training of trellising Tomato Plant, (b) Pruning and training of trellising Sweet Pepper Plant, (c) Training and Pruning of trellising cucumber plant, (d) Rose plant pruned (yearly) to basic frame and subsequent pruning at the time of harvesting.

High-density long-stem plants inhibit branching and grow the stems much taller (sometimes upto 2 m height). These are therefore supported by wire.

Carnation: After planting, the plant is to be clipped at a height of 1.5 m and the picking will be started from November. Necessary de-topping or de-budding is done to give a proper form to each plant and to regulate the quality of flower. The carnation grows on a trellis system with three to four horizontal lattices to keep the flowering stem erect.

Gerbera: No pruning as such, but picking of leaves should start at the age of 1 & $^1/_2$ year and the older leaves are picked as and when required. From 2nd year the plants require leaf pruning as well as pruning of rhizomes, depending upon the size and quality of cluster, in order to restart the cultivation afresh in an airy situation. However, old and dry leaves should be removed completely and regularly.

Chrysanthemum: (I) Pinching - Terminal buds of pot chrysanthemums are pinched to develop lateral branches and increase the number of flowers on the plant. Before pinching, several requirements must be met: a) the plant must have a root system that has reached the bottom and sides of the pot, and b) the development of 1½ - 1¾ inch of new growth on the cuttings, generally 12-14 days after planting. A soft pinch is used, removing about ½ inch of the stem tip, allowing 6-8 nodes to remain on the cutting. Hard pinches are not desirable because it reduces the number of breaks. (2) Debudding - For cut flower production, where a single stem is required, removal of all other buds, except the apical bud has to be done. (3) Training - Training of chrysanthemum is classified based on the handling of cultivars during production. These are of three types, (a) Standard - these types are usually grown single stem with all the lateral flower buds removed to develop one large, terminal flower head. This is usually used for cut flower production, (b) Disbud - these types are usually grown multi-stem (plants are pinched as rooted cuttings) with the lateral flower buds removed to develop one large, terminal flower head on each lateral. This is usually used for pot crop production, (c) Sprays - these types are usually grown multi-stem with only the terminal flower bud removed to allow all lateral flower buds to flower. This is usually used for pot crop production.

Training, pruning, of trellising Tomato

11. Trellising of tomato in fan-pad greenhouse   12. Trellising Egg plant in bamboo greenhouse

13. Trellising Capsicum in bamboo greenhouse   14. Tomato in bamboo greenhouse

7.8.4. Chemical growth regulation

Chemicals used for stimulating growth are derived from a group of compounds known as 'hormones'. Plant hormones are bio-chemicals produced in different parts of plant and affect the growth mechanism in different ways. Auxins and Gibberellins are two groups of hormones used by the greenhouse growers.

Auxin: Auxins are chemicals that promote growth at the apical parts of the plant. Indole-3-acetic acid (IAA) is a major auxin produced in plant. Auxins are produced in developing buds and young leaves and are transported downward (towards crown of the plant).

Auxins can be produced synthetically and are used for promoting rooting of cuttings. These are Indole butyric acid (IBA), Indole propionic acid (IPA) and Napthalene acetic acid (NAA). IBA and NAA are often found in combination as commercial product.

Auxins promote both shoots and roots simultaneously as 'tropistic' growth movement. Shoots grow towards the light source because auxin is believed to be inactivated by light hence inducing better growth in the darker side of the plant. Auxins also inhibit lateral shoot development. When top of the main shoot of a plant is removed, the source of auxin is lost and the lateral shoots are free to develop.

Gibberellins: A number of this hormone has been developed from the fungus species *'Gibberella* which commonly attacks rice plants and causes them to grow tall and threadlike. Gibberellins promote growth, but unlike auxins the promotion is uniform throughout the plant tissue. The growth stimulation occurs through increase in cell elongation. Forms of Gibberellic Acid (GA) have been used in different flower crops in proper doses to increase the size and quality of flower. However higher doses carry an adverse effect. GA inhibits root formation on leave and stem cuttings.

7.9. Harvest and Post Harvest Management of Different Greenhouse Crops

Though it is not directly connected with crop-husbandry, but a very important aspect of greenhouse cultivation. Particularly the harvesting stage and time is crucial for the crop. In general, flowers and vegetables are harvested by hand in the morning or in the late evening.

Gerbera is harvested when 2-3 whorls of stamens have entirely been developed. Plucking is done in morning or late evening. It is plucked from the plant to avoid any sort of cutting. After harvesting, the stocks are dipped in fresh water in a bucket and carried to the pre-cool unit avoiding any jerking. For that purpose an electric vehicle will be used. After reaching pre-cooling unit heel of the stock is cut 3-4cm and put into cold water added with sodium hypochlorite or citric acid + ascorbic acid for 4 hours. After 4 hours the individual flowers are put in a small polythene bag to protect the flower and then put into the box in bundle of 10 sticks (See photo at the end of this chapter). The boxes are then put into the cool-room at 15-16°C until it is loaded to the insulated A.C. van where the same temperature will be maintained.

In case of orchid and Anthurium almost similar pattern will be followed but with more precision and care. In case of Orchid the cut portion of the spike of each and every flower will be inserted into a tiny tube like container having water with or without preservatives (See photo at the end of this chapter). Otherwise that part will be wrapped with moist cloth or any such material. Orchid will be graded according to the length of the spike and no. of open flowers. 30 cm with 4-5 flowers, 40cm with 6-8 flowers, and 45cm with 8-10 flowers are graded as small, medium, and large respectively.

In case of Anthurium the market quality of flower is dependent on length of stock and diameter of the spadix. It is generally stored in 18-20°C. Anthurium will be packed in 40x16x5 cardboard flats that are padded with shredded newspaper and wrapped with insulation (See photo at the end of this chapter). In respect of gradation 5-73 spadix + 163 stem length, 4-53 spadix + 143 stem length, and 3-43 spadix + 123 stem length are called large, medium, and small respectively.

In case of vegetables, after plucking it will be put into the vegetable crate and transported to grading & packing house. After removal of field-heat at pre-cooling chamber it will be washed properly and dried accordingly. After drying it will be packed with shrink-wrap (expandable thin plastics) of desired weight and put into the plastic crate (See photo at the end of this chapter). Then the crate is put in the cool-room until it is loaded to the insulated AC van.

Lettuce, coriander leaf, and any other leafy vegetables will be harvested in two ways, 1st the whole plant harvesting and another is harvesting of leaves keeping the plant intact for next crop. The choice will depend on the variety, market, and season. After harvesting it will also follow the same post-harvest procedure as described above, obviously with some modification depending upon the crop.

Greading of capsicum and gerbera

Packaging technique for vegetables

Dendrobium flower   Water Capsule for Orchid cut flower

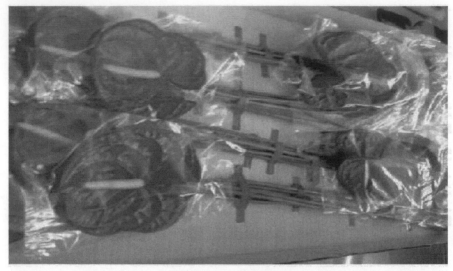

Anthurium ready to ship

Gerbera ready to ship

8

Tips for Cultivation of Some Important Greenhouse Crops

8.1. Vegetables

8.1.1. Tomato *(Lycopsarsicon Esculentum)*

Temperature: A warm climate crop can tolerate high temperature but cannot withstand frost. Minimum temperature for germination of seed is 8 to 10°C. Night temperature for fruit setting is 12 to 22°C. Critical night temperature for fruit setting is below 12°C. Critical day temperature for fruit setting is above 32°C. If the greenhouse can maintain the temperature as mentioned above then the non-terminal or indeterminate tomato crop can be grown for 10 to 12 month duration.

Variety: Indeterminate (producing flowers and fruits continuously along the stem as it grows) hybrid varieties resistant to nematodes or *fusarium* or *verticilium* wilt disease may be selected.

Beefsteak varieties	FA-574 & 180 and 514
Big-fruited varieties	Naveen, Akra Vardan & Visal, Nun-7711, NS-646, GS-600, Delphi, Astona, Comos, Shanmon, FA-189 & 179 etc.
Cluster type varieties	HA-646, FA-556 & 521.
Cherry tomato varieties	T-56, NS Cherry-1 & 2, BR.-124, HA-818.

Planting: 28-30 days old seedling will be planted with 60 X 50cm spacing. Each bed shall accommodate two rows only. The top 30 cm soil shall contain 30 % of organic matter. This layout gives a total of minimum 21,000 plants/ha.

Irrigation: Drip irrigation lines shall be laid before transplanting with laterals of 20mm diameter plugged with drippers having discharge capacity 2lit./hr at an interval of 50 cm. Two laterals per bed with a spacing of 60 cm is recommended.

Greenhouse tomato in soil-less culture

Fertigation: Fertilizer solution of 5: 3:6 ratio of N:P:K can be applied through drip @ 5 to 8 liter solution / m^3 of water according to the growth stage and season. It can be done @ 6 m^3 / 1000m^2 area at an interval of 6 to 8 days during September to November. Then the interval may be increased to 10-12 days and then further to 15 days during May- June. The other nutrients including micro-nutrients may be supplied through drip or directly into the soil as per requirement.

Training and Pruning: Due to indeterminate habit of growth, tomato crop in greenhouse needs regular training and pruning right from the few days after transplanting. Single stem shall be retained by removing all the side shoots or suckers, which develops between leaf petiole and stem. Plants are supported by plastic or binder twine, loosely anchored to the base of the plant with the help of clip or loop. These go vertically upwards and are attached to the strong load bearing support wire (11 to 12 gauge) running along the length of the entire row and fixed 8 to 9 feet above the ground. This twine shall be wrapped clock-wise around the stem as it develops with one complete swirl every three leaves. As fruits mature on the lower parts of the stem, the older leaves below the lowest mature fruit cluster shall be pinched off. When the stem reaches overhead load bearing wire, the twine is untied to allow dropping down of the vine/stem along with twine at least 2 to 3 feet at every 15 to 20 days. Fruit pruning is sometime practiced to maintain the size and shape of the fruits.

Other Operations: Inside air circulation through air circulating fans is essential, particularly from 10.00 AM to 1.00 PM and 2.00 to 3.00 PM, which induce pollination. Though tomato is a self- pollinated crop but aided pollination is needed for greenhouse tomato due to limited air movement and higher humidity.

8.1.2. Sweet pepper (*Capsicum annnum L.*)

Climate: Optimum night temperature for quality fruit development is 18-21°C. Night temperature below 16°C will affect the growth and yield adversely. Sweet pepper can tolerate day temperature above 30°C as long as night temperature is within 21-24°C. This crop is insensitive to photoperiod and humidity. However, hot climate with dry or wet weather results in poor fruit setting.

Variety: Two types of sweet peppers are cultivated in greenhouse condition viz. Red and Yellow. The commercial greenhouse varieties are; Red- Heera, Bharat, Indira, Bombay, Jamini, Pusa Deepti, Nun-3019 etc. Yellow-Tanvi, Boyaton, Orobellee, Golden Summer, Nun- 3020 etc.

Planting: The seedling, after spraying of systemic pesticide like 'confider' or 'metasystox' @ 0.5 ml / liter of water, shall be planted with spacing of 60 X 30 cm. in the beds. 4200 to 4300 seedlings are required for 1000 m^2 size greenhouse with this spacing. The transplanting shall preferably be done in the evening. The night temperature should be maintained at 21^0C.

Irrigation: Sweet pepper in greenhouse shall be irrigated through drip irrigation system. Two laterals will be laid per bed 60 cm apart, before transplanting. The laterals of 20mm diameter contain drippers along the laterals, spaced 30 cm in between, having discharge capacity of 2lit./hr. However, irrigation schedule @ 2-2.5, 2.5-3 and 2-2.5 m^3/ 1000m^2/day may be followed in stages up to fruit setting, from fruit setting to first harvest and first harvest to seven days before last harvest respectively.

Fertigation: In general N:P:K solution in the ratio of 5:3:6 is prepared for application in the sweet pepper crop. The schedule of fertigation of NPK at different stages is given below:

Nutrient	Up to fruit setting	Fruit set to 1st harvest	After1st harvest
N (ppm/m^3 of water)	80 to 100	120 to 150	100 to 120
P (ppm/m^3 of water)	50 to 60	75 to 100	50 to 60
K (ppm/m^3 of water)	100 to 120	120 to 150	100 to 120

Training & Pruning: The first terminal bud/flower, developed at the top of a single stem of the young plant having 9-13 leaves, is to be removed just after appearance. From that point the main stem will be divided into two or more branches. Two healthy branches are maintained on each plant after pruning or pinching out the other branches. Then all side- shoots, leaves and flowers are to be removed from each branch leaving two leaves and one flower in each internodes. Three leaves are left on side branches when the plant needs extra leaf area for more photosynthesis or to shade the fruits.

Capsicum in bamboo greenhouse

These two branches of each plant are trained upon strings fixed with the main wire running along the row at 8-9 feet height. The branches are either loosely trellised or bound around the strings. For each plant two strings are used and are fixed loosely using clip or other means with the main stem of each plant. Training and pruning shall be done every three or even two weeks. Crops can grow up to 8-9 feet in height in 9-10 months period.

Sweet pepper can hold fruits up to a certain extent as excess fruit load may cause decaying of roots.

Harvesting: It requires 60 days to obtain mature green coloured fruit and to obtain ripe yellow or orange coloured fruit another 3-4 weeks time is required.

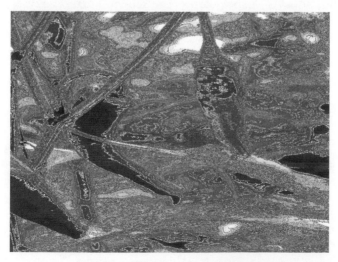

Kashmiri chilli in bamboo greenhouse

8.1.3. Cucumber (*Cucumis* spp.)

Temperature: Cucumber is much susceptible to temperature. Greenhouse cucumber is more sensitive to minimum temperature, lower than 18°C, for sustained production. Prolonged high temperature, above 35°C, affects adversely the quality and quantity of fruit production. The growth rate of cucumber depends on the average temperature of 24 hours, ideally 28°C, with a minimum fluctuation between day and night temperature.

Variety: For greenhouse cultivation, special type of cucumber varieties is to be used. These varieties are gynoecious and set fruits parthenocarpically i.e. fruit development takes place without pollination. Actually, natural cross-pollination results in bitter fruits. So for such varieties the greenhouse shall be protected from pollinating agents like bees etc. The parthenocarpic gynoecious varieties are of two types; producing only female flowers or producing mostly female flowers along with some male flowers. Satis, Alamir, Nun-9729, Kian are the parthenocarpic varieties available in India. However, monoecious varieties that produce both male and female flowers and need pollination can be grown in greenhouse without insect proof arrangement or having proper pollination management. Japanese long green, Pusa sanyog, Priya and Poinsett are some monoecious varieties that can be grown inside the greenhouse.

Planting: As it grows vigorously in greenhouse with large leaves and requires plenty of sunlight, the spacing between rows are kept at 1.4 to 1.5 m i.e. one row per bed with plant distance of 30cm. However, depending on varieties and better light availability, particularly in summer, the row may be placed 60 cm apart with plant to plant distance of 30-45 cm.

In greenhouse, cucumber can be grown round the year, in three growing cycles. The first cycle may be planted in January and will be picked from February to the end of April. The second cycle shall be planted in May and picked within a short period of 60 days (June to July). And the third cycle shall be planted in July or August and picking shall be continued up to late December. Alternatively two cycles may be grown in a year if the first cycle continues to produce well and yield satisfactorily up to July.

Trellising Cucumber in greenhouse

Irrigation: Watering of cucumber plants is to be done through well laid out drip irrigation system depending upon the spacing between rows and plants. Continuous water supply is essential for cucumber. In summer, irrigation is to be given @ 3.4 m³ and 3.5 m³ per 1000m²/day up to 1st picking and after 1st picking respectively. In winter these figures will be reduced to 2-2.5 and 2.5 -3 respectively.

Fertigation: Continuous and adequate supply of nutrients, particularly in peak fruit production periods, is essential for parthenocarpic cucumber. However the fertigation schedule may be as follows:

Nutrients	Up to 1st picking	After 1st picking
N (ppm/m³ of water)	80 to 100	100 to 150
P (ppm/m³ of water)	60 to 70	80 to 100
K (ppm/m³ of water)	100 to 120	120 to 150
Ca (ppm/m³ of water) if required	30 to 40	40 to 50

Training & Pruning: The plants are trained upward along a twine so that the main stem is allowed to climb to the overhead horizontal wire/steel cable running 8-9 feet above the ground. As the stems develop and cross the overhead wire it is again trained to downward direction.

No fruit shall be allowed up to 1.5-2 feet above the ground. Weak and unproductive laterals shall be removed. In case of excess fruiting thinning of fruit shall be done. Multiple fruit occurrence in axils should be thinned to one.

8.1.4. Eggplant or brinjal *(Solarium melongena)*

Climate: Eggplant is a warm climate crop that is very susceptible to frost or very cold night temperature. Prolonged warm period with a mean temperature of 20-30°C is most favourable. It requires a relatively long growing season to produce profitable yields. It can also sustain high humidity during growing period.

Egg plant in bamboo greenhouse

Variety: A wider range of shapes, sizes and colours is grown in India and some other parts of Asia. Larger varieties weighing up to a kilogram grow in the region between the Ganges and Yamuna rivers of India. Oval or elongated oval-shaped and black-skinned cultivars include: 'Harris Special Hibush', 'Burpee Hybrid', 'Black Magic\ 'Classic', 'Dusky', and 'Black Beauty'. Traditional, white-skinned, oval-shaped cultivars include 'Casper' and 'Easter Egg'.

The most widely cultivated varieties in Europe and North America are elongated ovoid, 12-25 cm long (4 1/2 to 9 in) and 6-9 cm broad (2 to 4 in) with a dark purple skin. Long, slim cultivars with purple-black skin include: 'Little Fingers', 'Pingtung Long', 'Ichiban' and 'Tycoon'; with green skin: 'Louisiana Long Green' and 'Thai (Long) Green'; with white skin: 'Dourga'. Bicoloured cultivars with color gradient include: 'Rosa Bianca', and 'Violetta di Firenze'. Bicoloured cultivars with striping include: 'Listada de Gandia' and 'Udumalapet'.

Planting: A well-drained sandy loam or loam soil, fairly high (10 to 30 % of soil) in organic matter with pH of 6.0 to 6.5 is best for planting. For planting use only strong and healthy transplants of 6 to 8 inches height. Commonly plants may be spaced 65 cm apart in rows and the rows are spaced 1 .2 m apart in beds. This system can accommodate about 6000 plants per acre in open field or in greenhouse. This density can also be achieved by planting the transplants in single row on 1.6 m wide beds with spacing of 40 cm along the row.

Irrigation & fertigation: Normally drip irrigation system with one emitter (2 lph) per plant is deployed with a total supply of 10 mm/day/plant in the peak season. This can be achieved by operating the system 12-14 hours per day.

Normally all required organic manure, micronutrient, and 50 % of phosphorus might be applied at the time of bed preparation. The rest 50 % of phosphorus may be applied directly to bed after 4 -6 months after transplanting. Just after transplanting, the required amount of nitrogen and potash will be supplied through drip system on regular basis throughout the life span of the crop.

Training & pruning: The plants shall be grown on trellises almost in the same way as in sweet pepper. The plants will be pruned in such a way that only two best-grown branches are allowed to grow on vertical wires. However, trimming of lower leaves are not done in eggplants, which is common in other trellised greenhouse crops.

Harvesting and Storage: Sharp knife or small pruning shears are used for harvesting. Harvested at least once a week, preferably twice a week, before flesh becomes tough and seeds begin to harden. The market usually prefers fruits 4 to 6 inches in diameter. Fruits are usually sold in bushel baskets, crates or cardboard containers.

Eggplant loses water and quality when field heat is not removed quickly. In storage room forced air and/cooling works well for eggplant. They should be stored at 45 to 50°F to avoid chilling. Relative humidity should be at least 90 percent. They should not be stored for longer than 10 to 14 days, before retailing. Long-term storage in cold-chamber results chilling injury (surface scald or bronzing and pitting) to the fruits.

8.2. Flowers

8.2.1. Rose (*Rosa* spp.)

Temperature & Light: Temperature range from 15°C to 28°C is ideal. The difference between day and night temperature should be above 5°C. It can be grown below 15°C but interval between flushes become longer. Roses can be grown above 30°C with high humidity and slow evaporation rate. Plenty of light is essential for rose. Shading is sometimes required to prevent scorching of petal.

Variety: (l) Hybrid Tea, long stem (125 cm) and large flower (4 cm) - Sonia, Vivaldi, Tineke, Melody. Darling, Only love etc. (2) Floribunda - shorter stem (60 cm) and small flower (2.5 cm) - Mercedes, Frisco, Jaguar, Kiss, Florence etc.

Planting: High density planting @ 70,000 plants/ha with 30 x 17 cm spacing may be recommended. Planting may be done during February to April or July to September.

Irrigation: It requires at least 6 mm/day i.e. about 60 cm/ha/day through drip irrigation system.

Fertigation: Nitrogen and potassium @ 200 ppm is to be applied twice a week for 7 months along with drip irrigation system. Phosphorus and organic matter @ 1.8 kg and 15 kg /m^3 soil (up to 30 cm) respectively may be applied directly into the soil. Other nutrients including micronutrients shall be applied to the soil as per requirement.

Pruning and Training: From initial stages only l to 2 main stems are allowed to grow to form the basic structural frame by proper training. The frame should not exceed the spacing area allotted to an individual plant. Every year after the production season these main stems are cut back leaving 4 to 5 nodes. Last year's flower producing branches should also be pruned out. All weak stems shall be removed and the wayward stems shall be redirected by proper pruning. For older plants, roots may be pruned to promote the development of new young roots. However, in case of high density planting the plants are uprooted, for replacement, every 2 to 3 years. Apart from yearly pruning, the plants regularly get pruned automatically while lowers are harvested with long stalk having 4 to 6 full grown leaves. In general the flowering stalks are cut above the node with last leaf having 5 leaflets. This helps to grow new strong flowering stem from petioles of the leaves below the harvesting cut.

High-density long-stem plants inhibit branching and grow the stems much taller (some time up to 2 m in height). These are therefore supported by wire.

Rose in large fan-pad greenhouse

8.2.2. Carnation (*Dianthus* spp.)

Temperature & light: Most carnation varieties are photoperiod insensitive. High light intensity with a 12 hr. day length may produce top quality flower. The ideal temperature requirement is about I0°C at night and 23°C during day.

Variety: Standard carnation - Kletonwi, Sarina Lonseva, Candy klekopi, Candy klemaxi, Sandrose, Manon Korsa, Tanga Lotarion etc. Spray carnation.- Media, Silvery pink, Lior Aroiler, Salmony stasalm, Binaca, Starlight hilstaretc.

Tissue culture plant material or disease free cuttings of desired variety may be used for planting.

Planting: Carnation may be planted in 20 cm raised bed of 1 m width with a 0.5 m path between the beds. The soil based root medium should have sufficient sand particles, for proper aeration, and also have high organic matter. Planting may be done in August to October with a spacing of 20 cm x 15 cm. The target of layout is to accommodate 30-35 plant/m^2.

Irrigation: Irrigation shall be done through drip system. 20 lit of water per m^2 is required to apply twice a week @ 2 Iit/hr/dipper.

Fertigation and manuring: Nitrogen and Potassium may be applied through drip @ 200 ppm at 15 days interval. 38 ppm Boron may also be applied through the drip system along with nitrogen and potassium. Phosphate and organic manure may be applied directly into the soil @ 700 kg/hr/yr. (in two splits) and 10 kg/m²/yr respectively. The other nutrients are to be applied into the soil as per requirement.

Pruning and other practices: After planting the plant is to be clipped at a height of 1.5 m and the picking will be started from November. Necessary de-topping or debudding is done to give a proper form to each plant and to regulate the quality of flower. The carnation grows on a trellis system with three to four horizontal lattices to keep the flowering stem erect.

Carnetion growing in Greenhouse

8.2.3. Crysanthemum (*Dendranthema grandiflora*)

Light: Chrysanthemums require the maximum light intensity (5000-6000 ft.ca.) available as long as temperature can be controlled. Growers in the hot area find it necessary to apply a 25-30% shade during the summer to control heat. The chrysanthemum is a qualitative short-day plant in respect to flowering with temperature modifying the photoperiodic response. Plants flower when the day-length is shorter than the critical day-length and grow vegetatively when the day-length is longer than the critical day-length. However, chrysanthemums have two critical photoperiods, one for floral initiation and a different one for flower development. However, there is no single critical photoperiod because it can vary depending on the cultivar and temperature. Supplemental light from high intensity discharge (HID) lamps to improve vegetative growth is often used during low-light period. Lighting is supplied for 16 to 24 hours per day. Artificial short days must be created for chrysanthemum during long-day times of the year to ensure flower initiation and development. This is accomplished by pulling opaque material (black cloth) over the plants for 12-15 hours per day.

Chrysanthemum in large hi-end greenhouse

Carbon dioxide: Chrysanthemums benefit from the application of supplemental carbon dioxide during the vegetative period with increased dry weight, increased lateral branching, larger leaf area, and shorter production times. Growers, who utilize supplemental carbon dioxide, apply it at the rate 800-1000 ppm mostly during propagation and the early vegetative period.

Temperature: The temperature generally declines from the beginning of short days which helps to open the flower or flowering. The optimum temperatures in between the starting of short days and flower bud initiation are 20-22°C night and day. Temperatures above and below 20-22°C will delay flowering. After visible bud, optimum temperatures decrease to 18°C to about the 'showing color' stage and to 15.5°C for the last two weeks. Temperatures below 15.5°C will delay flower development. The optimum temperatures for flower development are lower than those of many other crops. Therefore, chrysanthemum crops should have a dedicated finishing area for temperature control.

Variety: Chrysanthemum is available in a wide range of colors, flower types, and plant sizes. Of these, some are more suited for cut flower production and others for outdoor planting. However, about 100 cultivars are best suited and are widely grown nowadays as flowering pot plants.

Planting: In general, under greenhouse condition, pots are used for growing chrysanthemum. Rooted cuttings, purchased or produced, should be graded by size so that individual pots/beds receive the same sized cuttings. Unequal cutting size in the same pot/bed inevitably results in uneven growth at finishing. Cuttings may be graded into: (1) short - thin stem diameter and not well rooted, (2) average - medium stem diameter and fairly well rooted, and (3) tall - thick stem diameter and well rooted. However, 6" pots require 4 to 5 cuttings per pot depending upon the growing season. Bigger size pots (8") require more number of cuttings. 4" pots always contain one cutting. Cutting should be planted shallow with the roots just covered with medium. In 6" pots they are planted as close to the pot rim as possible and leaning out from the rim. This allows the maximum space for lateral shoot development and light penetration. Cuttings should be planted in a moist medium and watered twice soon after planting. The first watering should be clear water followed by a fertilizer solution.

Irrigation: Chrysanthemums require plentiful of water and fertilization. The media should be moist, or allowed to dry only slightly between two watering. Sufficient water should be applied per watering to completely saturate the medium plus 10-15% leachate. On an average, pot size of 4", 6" and 8" require 13, 31, and 55 gms of water per irrigation. Manual watering cannot provide required degree of control over soil moisture for quality pot culture. Most of the good growers have their crops on some form of automatic watering system like micro-tube system or other drip irrigation system.

Fertigation: Chrysanthemums require large quantities of fertilizer during the vegetative stage of production, especially nitrogen and potassium. In case of pot culture many growers water chrysanthemums using 300-400 ppm N using a balanced fertilizer. This high rate of nitrogen may be applied at each watering until the roots reach the bottom of the pot, then the fertilizer rate is reduced to 200-250 ppm N on a constant liquid feed basis. Fertilizers high in ammonium (NH_4) in the vegetative stage during the warmer periods of the year induce rapid growth. Generally fertilizer rates should be reduced to about 125-150 ppm N after flower buds become visible. Several studies have shown that post-production keeping quality is improved when fertilization is totally stopped 3-4 weeks before the end of vegetative growth or about the time of 'dis-budding'. However, benefits have been observed on pot chrysanthemums when liquid fertilization is combined with a slow-release fertilizer.

Pruning and training: (1) 'Pinching' - Terminal buds of pot chrysanthemums are pinched to develop lateral branches and increase the number of flowers of the plant. Before pinching, several requirements must be met: (a) the plant must have a root system that has reached the bottom and sides of the pot, and (b) the development of 1 ½ -1 ¾ inch of new growth on the cuttings, generally 12-14 days after planting. A soft pinch is used, removing about ½ inch of the stem tip, allowing 6-8 nodes to remain on the cutting. Hard pinches are not desirable because it reduces the number of breaks. (2) 'Debudding' - For cut flower production, where a single stem is required, removal of all other buds, except the apical bud is done. (3) Training - Training of chrysanthemum is classified on the basis of handling of cultivars during production.

There are of three types of training; (a) Standard - these types are usually grown single stem with all the lateral flower buds removed to develop one large, terminal flower head. This is usually used for cut flower production, (b) Disbud - these types are usually grown multi stem (plants are pinched at the stage of rooted cuttings) with the lateral flower buds removed to develop one large, terminal flower head on each lateral. This is usually used for pot crop production. (c) Sprays - these types are usually grown multi-stem with only the terminal flower bud removed to allow all lateral flower buds to flower. This is usually used for pot crop production.

8.2.4. Anthurium *(Anthurium Andrianam)*

Temperature & Light: Anthurium thrives best in a day temperature of 25° to 28°C and night temperature of 18° to 20°C. It is a shade loving plant and the shade requirement of the plant ranges from 60 to 80% of the full sunlight. Insufficient shading causes damage to the plant. It also requires 70 to 80% relative humidity.

Hydroponics for Anthurium   Anthurium in high-tech greenhouse

Variety: Red - Ozaki, Kaumana, Kozohara, Mirjan etc.
Orange: Nitta, Sun burst, Favoriet etc.
Pink: Abe, Avo-Anneke
White: Haga white, Manoa Mist, Chameleon etc.

Planting: Before planting the root medium is required to be prepared meticulously with leaf mould, FYM, large piece of semi decomposed plant parts (twig, bark, shell etc), charcoal, sand, small brick bats/porous stones with a ratio of 10 : 5 : 1 : 1 : 3 : 1. No soil is required. Planting may be done with 30 cm x 30 cm spacing.

Irrigation: Judicious and regular watering is essential for good growth. Sprinkling or misting beneath the canopy is the best way of irrigation. However, microclimate of greenhouse, size and growth stage of plant are to be taken into account while watering.

Fertilizer application: 2g NPK @ 13 : 20 : 20 per month per plant may be applied into the root medium as solution in such a way that the root media get soaked with the nutrient solution. To avoid calcium deficiency (which is common) $CaCO_3$ @ 4 gm/plant/year may be applied.

Treatment of plants with IBA or Ethephon induces adventitious buds.

8.2.5. Gerbera (*Gerbera jamesonii*)

Temperature, Humidity & Light: After planting, the young plants require higher temperature at night (20 to 22°C) for fast rooting and initial growth. After establishment, the plants require a night temperature of 12 to 16°C and day temperature of 20° to 25°C for good flower production. 80 to 35% relative humidity is optimal for Gerbera. As a plant, gerbera grows well in sufficient sunlight. However, in strong solar radiation the plant is forced to high evapo-transpiration, which causes temporary wilting due to insufficient water uptake by the roots. Temporary wilting can be avoided by providing 50% shading at strong solar radiation period of the day with a movable shading screen.

Gerbera in fan-pad greenhouse

Variety: Ibiza (double, cream coloured with black centre), Rosabella (semi double, light pink), gold spot (double, yellow colour with black centre), etc.

Planting: Gerbera shall be planted in bed with high aeration and permeability. Mixing of sand, coconut coir pith, paddy husk with soil can provide the desired aeration and permeability of the soil root medium. The planting may be done at a distance of 30 to 40 cm from row to row and 25-30 cm from plant to plant, which results in 6 to 7 plants per m^2 of bed area. The plant shall not be planted deep into the soil. The crown region shall be placed above the ground surface.

Irrigation: About 500 to 700 ml of water /plant/day is required to irrigate through drip irrigation system, depending upon the season and stage of the crop. However, in the initial stages up to the first flowering, irrigation can be done through sprinkler system. In cloudy days irrigation should be given in restricted manner.

Manuring and fertilizer application: Application of organic matter (FYM) @ 10 to 15 kg/m^2 of bed area may be applied at the time of bed preparation. Then application of NPK @ 15: 10:30 gms /m^2/month shall be given either through drip irrigation system or directly into the soil in two splits at 15 days interval around the plants. Before flowering starts i.e. first 3 months, the dose will be 10 : 15 : 20 gm NPK/m²/Month. Application of micronutrients like boron, calcium, manganese and copper @ 0.15%* (1.5 ml/lit. of water) each may be applied per month.

Pruning: Not pruning, but picking of leaves should start at the age of 1 & ¹/₂ year and the older leaves are picked as and when required. From 2ⁿᵈ year the plants demand leaf picking as well as pruning of rhizomes depending upon the size and quality of cluster in order to start cultivation afresh with an airy situation. However, too many picking at a time is not advisable. Old dry leaves should be removed completely and regularly.

8.2.6. Orchids (*Dendrobium/Vanda/Cattlea/Phalaeonopsis*)

Here we will only consider the commercial Epiphytes like Cattlea, Dendrobium, Vanda and Phalaeonopsis.

Temperature, light and Humidity: 12°C minimum and 32°C maximum temperature is suitable for the said orchids to thrive well. The required light intensity is 2000 to 6000 foot candals thus requiring shade throughout its growing period. 50% to 75% shade is required depending upon the age of the plant and intensity of solar radiation. High humidity with a range of 60% to 90% relative humidity is required for proper growth of orchids.

Variety: Dendrobium - Mame Pompadour, Louis Bleriot, Jacquelline, Thomas, VH 44 etc. Cattleya - Small world, Brabantaiae, Angel walker, Easter etc. Vanda-Miss Joaquin, Rothsehildiana etc.

Planting: These Epiphytes are generally grown in big pots or vessels made up of plastic or earth. The root mixture shall contain the materials like charcoal, Sphagnum mosses, tree fern fiber, fine bark, coconut husk, wood cube, broken bricks/jelly stone, other aggregates (solite/holite) etc. in such a combination that the media shall provide very good drainage, aeration and support.

17. Dendrobium in naturally ventilated greenhouse

Irrigation: Watering can be done manually or through sprinkler 1 to 3 times daily or even at larger intervals i.e. once in two days depending upon the species, age of plant, container, media mixture and climate. The amount of water is up to the saturation of the components of the medium. The water should not contain soluble salt more than 700 ppm.

Fertilizer application: For commercial production of cut flower orchids, plants are fertilized with NPK and selective organic manures. NPK may be given @ 19 : 19 : 19 or 20 : 20 : 20 supplemented with trace elements at a concentration 0.2% in summer and 0.1 % in winter after every watering in the morning. The pH should be adjusted between 5.5 and 6.5 once in a week in growing season and lean season respectively.

8.3. Fruit

8.3.1. Strawberry (*Rosaceae Fragaria*)

Key point: Flowers of most cultivers are bisexual and self-pollinated. However supplementary pollination by hand or by Bumble Bee (50 no. for 4000 plants) is essential for greenhouse cultivation.

Climate: Strawberry basically is a cool climate crop. The average growing temperature is about 35°C and 15°C at day and night respectively with 9 hour photoperiod. It requires necessary moist and ventilated condition inside the greenhouse. At the time of blossom it is better to have 2 to 3°C more temperature at night and somewhat dry environment.

Planting: Strawberry is normally cultivated in pots or in raised soil-based beds covered with straw. However, for commercial greenhouse cultivation three types

Strawberry in greenhouse

of growing systems are practiced e.g. (1) beds with straw mulch - the poorest method, (2) Pots or plastic bags placed on the ground or hanged 1.2 m above the ground, and (3) Troughs made up of plastic or other suitable material hanged 1.2 m above the ground.

Soil-less media is prepared from coconut-peat and perlite or any peat base substrate. The media may contain sand, charcoal dust and well-decomposed compost in required proportion. In general 4-5 plants/m² is planted in beds or in pots/bags placed on the ground. This plant density can be doubled or more if the hanging growing system is adopted where virtually no space of the greenhouse is left for path in between rows. The center-to-center spacing between rows may be 55 to 65 cm depending upon the system.

Irrigation: The plants are commonly irrigated through drip system with 10 cm spacing between emitters having 1 lph discharge capacity. During growing and flowering period irrigation is to be provided every day from 8 am to 5 pm with a schedule of 1 minute on and 90 minutes off.

Fertigation: Plants receive nutrient with every irrigation and each plant receives about 140 ml of nutrient solution per day. The best of such solution should contain the following ppm of macronutrients and micronutrients respectively. N:80, P:50, K:85, Ca:95, Mg:40, S:65, and Fe:2.4, Bo.0.6, Mn:0.4, Cu:0.1, Zn:0.2, Mo:0.03. The pH and EC will be 6-6.2 and 1.4-1.6 respectively.

Canopy management: Strawberry plant is creeper in nature; hence position of its fruits and leaves is a vital aspect for commercial production. The peripheral leaves and particularly the fruits are judicially trained to lean out from the pot/bag or raised bed.

Fruit Pruning: If more than eight to ten fruits are set in a plant, the smaller and imperfect ones should be removed. Apart from this, the dried and diseased leaves are also to be removed.

Plant Protection: For disease control, Bordeaux mixture in one to one thousand solution may be applied in growing stage. *Bacillus thurengiensis* var. israelensis may be applied every two weeks through the fertilizer injectors for controlling fungus gnats. To control powdery mildew suitable bio-fungicide (like *Tricoderma viridi, Ampleomyces quisqualis*) may be used.

As it is consumed fresh, biological control of different pests (aphids, spider mite and thrips) is always recommended. For that matter Ladybug beetle, parasitic Wasp, Big eyed bug, Insidiosus flower bug, predatory mite etc may be used to control the pests.

8.3.2. Grapes (*Vitis vinifera*)

Many people are under the impression that growing vines inside a greenhouse is invariably a non-viable proposition. But, with a bit of care in selecting the right variety, it is perfectly possible to have the crop of grapes in your greenhouse.

Varieties such as "Black Hamburgh" or "Buckland Sweetwater" are ideal for cooler growing conditions and of course, there are many options open to the warmer climate. However, very high temperature at the time of flowering and fruiting is detrimental to the crop. Thus, proper management should encourage dormant status of the crop in such yearly extreme climate (summer or winter).

Vines require open, free-draining soil to flourish and it is important to get this aspect of their cultivation right. Young plants will need support as they become established – stout canes being the usual choice – the new spring/new growths being trained to the support as they develop. Vines produce their flowers and then their fruit on each year's new growth, so promote a good crop; the side-shoots produced in the summer should be pruned in the following winter, cutting them back to their last bud – a method known as "spur" pruning.

The plants will need some attention over the year to keep them in peak condition and must be watered thoroughly to ensure that sufficient water reaches all of the roots and fed with a good quality fertiliser. As the fruits develop, they need to be thinned to allow each grape the space it needs to grow to a good size – an operation usually carried out with long scissors with pointed ends, working upwards from the bottom of the bunch and removing the smaller ones.

Unfortunately, vines suffer from a number of pests, including aphids, red spider mites and scale insects, so a careful look out needs to be kept – and prompt action taken – at the first sign of trouble. The fruits can also be prone to mildew. With care and a little effort, however, even the smallest greenhouse can provide a few bunches of home-grown grapes.

Grape in greenhouse

8.3.3 Guava (*Psidium Guajava*)

Guava can be opted for greenhouse cultivation, because they are self fertile, really tough, and easy pruned to keep it a small tree-shrub to accommodate as ultra high density population in a greenhouse.

There are a few suitable varieties that can be used for greenhouse. The two main strains are the apple shaped green and red, and the pear shaped golden and yellow. There is also the excellent cherry or mountain guava (*P. cattleianum*), which is evergreen with smaller leaves, hardy, and has purple red fruits which are delicious.

8.3.4. Cherries

Cherries make an excellent choice for greenhouse gardening, as they are susceptible to cold winter weather outside in cooler region and birds that pick on them. Cherries may be grown in a greenhouse to get a head start on the season, or extend their growing season to enjoy them in savory dishes long afterward.

Though it is a moderately big tree, but for greenhouse the dwarf varieties are to be chosen. Semi-dwarfing 'Gisela 5' and 'Tabel', reaching 3-4m (10-13ft), thus they are ideal as dwarf bush trees or for greenhouse. A recent and exciting development in sweet cherries is the dwarf self-pollinating "Stella." Furthermore it should be trained and pruned (yearly) properly to make it fit for greenhouse (See photo). Thus it can be used for ultrahigh density plantation in greenhouse.

For sweet cherries, more than one tree shall be grown as they are cross-pollinating. Dwarf sweet cherries can be planted 5 to 6 feet apart in double row system in greenhouse, where row to row distance may be 4.5 to 5 feet.

Cherry plantation technique for large greenhouse

Cherries do not like extreme heat, so several fans have to be kept for circulating air in the greenhouse if temperatures in the summer reach 100 degrees F. Cherries prefer cold temperatures in the winter, but the temperature shall be prevented from falling below freezing. Most of the sweet varieties do especially well in areas that have hot, dry summers, and a cooler winter. It needs about 400 chill hours and will fruit in next 6 to 9 days.

Sweet cherries fruit on one-year-old and older branches. Thus pruning should create a balance between older fruiting wood and younger replacement branches.

Formative pruning takes place in spring as the buds begin to open. The established trees are pruned from late July to the end August.

Annexure

Annexure : Conversion Formulas

Length

1 Inch = 25.4 millimeter = 2.54 centimeter = 0.0254 meter.
1 Foot = 12 inches = 30.48 centimeter = 304.8 millimeter.
10 Millimeter = 1 centimeter = 0.394 inches = 0.03283 foot.
1 Meter = 100 centimeter = 3.28 feet = 1.0936 yards.
1 Kilometer = 100 meter = 0.6214 mile = 1093.664 yards.

Area

1 Hectare = 2.471 acres = 10,000 sq meter = 11959.64 sq yards.
1 Acre = 0.405 hectares = 4840 sq yards = 4046 sq meter = 43560 sq feet
1 Sq feet = 0.1111 sq yards = 0.0929 sq meter.
1 Sq meter = 1.1960 sq yards = 10.764 sq feet.
1 Sq km − 0.386 sq mile = 100 hectares = 247.106 acres.

Volume

1 Liter = 1000 cubic centimeter (cc) = 0.0351 cubic feet = 0.22 gallon.
1 Cubic meter = 1000 liter = 35.1 cubic feet = 220 gallons = 1.3080 cu yards.
1 Cubic feet = 0.0283 cubic meter = 28.49 liters = 6.268 gallons.

Mass (weight)

1 Gram = 1000 mille gram = 0.0353 ounce.
1 Ounce = 0.625 pound = 28.35 gram = 437.5 grains.
1 Kilogram = 1000 gram = 2.2046 pound = 35.2736 ounce.
1 Ton = 10 quintals = 1000 kg = 2240 pounds.

Speed

1 mile / hour = 1.6 Kilometer / hour = 0.869 knots.
1 Kilometer/hour = 0.2778 meter/second = 3.288 feet/second.

Temperature

T^0 Centigrade = (5/9) (T^0 Fahrenheit + 40) − 40.
T^0 Fahrenheit = (9/5) (T^0 centigrade +40) − 40.

Pressure

1 Pound/sq inch (psi) = 702.7 kg/sq meter = 14.697 atm.
= 6.894757×10^3 pascal (Pa).
1 Pound / sq feet = 4.88 kg / sq meter = 47.88 Pa.
1 atmosphere (atm) = 1.0132 bar = 760 mm mercury = 101.325 kilo Pa.
1 bar = 0.987 atm = 1×10^6 dynes/sq cm = 10^5 Pa = 100 kilo Pa.

Illumination

1 lumen / sq meter = 1 Lux = 0.01 watt /sq meter.
1 Lumen / sq feet = 1 foot candle(fc) = 10.7639 lux.

Bibliography

A.M. Abd-Alla; S.M. Adam; A.F.Abou-Hadid and M.M. W. (1996). Awad Tomato production under different temperatures, irrigation and pruning regimes. *Ishs symposium on strategies for market oriented greenhouse production, March* 11-15, 1995 Alexandria-Egypt. *Acta Hort.*, 241-247.

A.M. Abdel-Mawgoud; S.O.El-Abd; A. F. Abou Hadid; T. C. Hsiao and S. M. Singer (1996). Effect of shade on the growth and yield of tomato plants. *ISHS symposium on strategies for market oriented greenhouse production, March 11-15, 1995 Alexandria-Egypt.Acta Hort.* (434) pp. 313-319 .

Abou-hadid, A.F. and Eissa, M.M., (1994). Daily air temperature and relative humidity regimes in relation to plastic houses and open field condition in Egypt. *Acta Horticulture,* 245, 113-118.

Abou-Hadid, A.F. (1990) Evaporation under protected cultivation in relation to seasonal changes in climatic factors. *Egypt. J. Hort.* 18 (1): 1-10.

Abou-Hadid, A.F., Maksoud M.A. and S. El-abd. (1986). Protected cultivation for winter production of tomato. *Acta Horticulture* (191), pp. 59-66.

Adams, P., (1994). Nutrition of greenhouse vegetables in NFT and hydroponics system, *Acta Horticulture,.* 361:245-257.

Adams, P. and L.C. Ho. 1993. Effects of environment on the uptake and distribution of calcium in tomato and on the incidence of blossom-end rot. *Plant and Soil,* 154:127-132.

Aikman,D.P.,A procedure for optimizing CO_2 enrichment of a glasshouse tomato Crop, *J. Agric. Engg. Research*, 63(2):171-183(1996).

Al-Arifi, A., T. Short and P. Ling. 1999. Influence of shading ratio, air velocity and evapotranspiration on greenhouse crop microclimate. *ASAE Paper* no. 99-4228, pp 17.

Al-Kayssi, A.W., (2002). Spatial variability of soil temperature under greenhouse conditions, *Renewable Energy*, 27:453-462.

Al-Shooshan, Ahmad, Ted H. Short, Robert McMahon, R. Peter Fynn. (1991). Evapotranspiration measurement and modeling of a greenhouse grown chrysanthemum crop. *ASAE Paper* no. 91-4043, pp 20.

Albright, L.D. (1994). Predicting greenhouse ventilating fan duty factors and operating costs. ASAE paper No. 944576. ASAE, 2950 Niles Road, St. Joseph, MI 49085-9659, USA. 20 pp.

Albright, L.D. (1997). Ventilation and shading for greenhouse cooling. *Proceedings of the International Seminar on Protected Cultivation in India*, December 18-19, Bangalore, India. pp. 17-24.

Albright, L.D. and H.I. Henderson. (1995). Efficient cooling and dehumidification technologies for the Cornell University Controlled Environment Agriculture (CEA) Complex. Final Report submitted to *NYSERDA*. 27 pp. with Appendices.

Arbel, A.,Yekutieli, O. and Barak, M., (1999). Performance of a fog system for cooling greenhouses, *J. Agric. Engg. Research*, 72(2):129-136.

Bailey, B.J. (1977). Heat conservation in glasshouses with aluminised thermal screens. Acta Hort. 76:275-278.

Bailey, B.J., (1981). The reduction of thermal radiation in glasshouses by thermal screen, *J.Agric. Engg. Research*, 26,215-222.

Baielle, A., Aries, F., Baille, M. and Laury, J.C. (1990). Influence of thermal screen Optical properties on heat losses and microclimate of greenhouse, *Acta Horticulture,*174:111-118.

Bakker, J.C., Effects of day and night humidities on yield and fruit quality of glasshouse tomatoes, *J. Hort. Sci.*, 65:23-331.

Baptista, F.J., Bailey, B.J., Randal, J.M. and Menses, J.F. (1999). Greenhouse ventilation rate, theory and measurement, , *J.Agric. Engg. Research*, 72:363-374.

Barral, J.R., Galimberti, Pablo, D., Barone, A., Lara, M.A., Integrated thermal improvements for greenhouse cultivation in the central part of Argentina, *Solar Energy*, 67(1-3):111-118.

Bartok, Jr. J.W. and Aldrick, R.A., (1984). Low cost solar collector for greenhouse water heating, *Acta Horticulture*, 2(184):771-774.

Bot, G.P.A. (1983). Greenhouse climate: from physical processes to a dynamic model. Ph. D. thesis, Agricultural university of Wageningen. 240pp.

Both, A.J. (1994). HID lighting in horticulture: a short review. Presented at the Greenhouse Systems, Automation, Culture and Environment (ACESYS I) Conference, July 20 - 22. New Brunswick, NJ. *Northeast Regional Agricultural Engineering Service Publication No. 72*. Riley-Robb Hall, Cornell University, Ithaca, NY 14853. pp. 208-222.

Boonen, C., O. Joniaux, K. Janssens, D. Berckmans, R. Lemeur and A. Kharoubi. (1999) Relationship between leaf temperature and the three-dimensional distribution of air temperature around a tomato plant. *ASAE Paper* no. 99-4126, pp 11.

Boulard, T. and A. Baille (1993). A simple greenhouse climate control model incorporating effects of ventilation and evaporative cooling. Ag. and For. Meteor. 65: 145-57.

Boulard, T.; A. Baille. (1995) Modeling of air exchange rate in a greenhouse equipped with continuous roof vents. *J. of Ag. Eng. Res*. 61:37-48.

Boulard, T. and B. Draoui. (1995) Natural ventilation of a greenhouse with continuous roof vents: measurements and data analysis. *J. of Agr. Eng. Res.* 61:27-36.

Boulard, T.; P. Feuilloley, and C. Kittas. (1997). Natural ventilation performance of six greenhouse and tunnel types. *J. of Ag. Eng. Res.* 67:249-266.

Boulard, T.; G. Papadakis, C. Kittas and M. Mermier. 1997. Air flow and associated sensible heat exchanges in a naturally ventilated greenhouse. *Ag. and For. Meteor.*, 88:111-119.

Briassoulis, D.,Waaijenberg, D., Grataud, J. and Elser B. Von, (1997). Mechanical properties of covering materials for greenhouse : Part 2 Quality assessment, *J. Agric. Engg. Research,* 67(2).

Buckley, R.A., Henley, R.W. and McConnell, D.B., (1993). Fan and Pad evaporative cooling system, *Florida Cooperative Extension Service*, University of Florida Circular,1135.

Buckman, H.O. and Brady, N.C. (1996). The Nature and Properties of Soils, *Eurasia Pub. House(p) Ltd.*,New Delhi.

Carvalho S.M.P., E. Heuvelink, R. Cascais, and O. Van Kooten, (2002) Effect of Day and Night Temperature on Internode and Stem Length in Chrysanthemum: Is Everything Explained by DIF? *Ann Bot.*, 90(1): 111–118.

Cebula S.,Black and transparent plastic mulches in greenhouse production of sweet pepper.Thermal conditions and vegetative growth of plants, *Folia Horticulture,*7(2):51-58(1995).

Challa H. and Schapendonk, (1984). Quantification of effect of light reduction in greenhouse on yield, *Acta Horticulture*, 2(184):501-510.

Chandra, P., Sirohi, P.S., Behera, T.K. and Singh A.K., (2000). Cultivating vegetables in polyhouse, *Indian Horticulture*.

Cohen, Y., M. Fuchs, V. Falkenflug, and S. Moreshet. (1988). Calibrated heat pulse method for determining water uptake in cotton. Agron J 80:398-402.

Coopper,P.I. and Fuller, R.J. (1983). A transient model of the interaction betweencrop environment and greenhouse structure for predicting crop yield and energy consumption, *J.Agric. Engg. Research*, 28:401-417.

Cotter, D.J and Walker, J.M. (1968). Strategy of relative humidity and temperature control in greenhouses, *Proc VIII National Agricultural Conf.*, Univ. of California, San Diego, USA.

Crockett, J.U., Greenhouse Gardening, Time Life Books, Alexandria, Virginia (1977)

Davidson, O.W. (1953). High pressure spray for greenhouse airconditioning. *New Jersey Flower Growers' Assoc. Bull.*, 3,4.

De Jong, T. (1995). Natural ventilation in large multi-span greenhouses. Ph. D. thesis, Agricultural university of Wageningen. 116 pp.

De Kreij, C. (1996). Interactive effects of air humidity, calcium and phosphate on blossom-end rot, leaf deformation, production and nutrient contents of tomato. J. Plant Nutrition 19:361-377.

De Graaf, R. and Van Den Ende, J., (1981). Transpiration and evapotranspiration of the glasshouse crops, *Acta Hort.*, 119:147-158.

Dilara, P.A. and Briassoulis, D., (2000). Degradation and stabilization of low intensity polyethylene films used as greenhouse covering materials, *J.Agric. Engg. Research,* 76:309-321.

Donahue, R.L., (1958). Soils – An Introduction to Soils and Plant Growth, *Prentice-Hall, Inc.*, New York.

Dooreenbose, J. and Pruitt Jr. W.O., (1977). Crop Water requirements, *FAO Irrigation and Drainage*, Paper 24, Rome.

Elly Nederhoff, (1997). The Best Way to Measure Humidity., *Commercial Grower,* Vol 52.

Elsner, B.von, Briassoulis,D., Waaijenberg, D., Mistriotis, A., Zubeltitz, Chr. Von, Gertraud, J., Russo, G. and Sauy-Cortes, R. (2000). Review of structural and functional characteristics of greenhouse in European union countries, Part I: Design Requirements, *J. Agric. Engg. Research*, 75(1):1-16.

Elsner, B.von, Briassoulis, D., Waaijenberg, D., Mistriotis, A., Zabeltitz, Chr. Von, Gertraud, J., Russo, G. and Sauy-Cortes, R., (2000). Review of structural and functional characteristics of greenhouse in European union countries, Part II: Typical Designs, *J. Agric. Engg. Research*, 75(2):111-126.

Fernandez, J.E. and B.J. Bailey. (1992). Measurement and prediction of greenhouse ventilation rates. *Ag. and for. Meteor.* 58:229-245.

Ferentinos, K.P., Albright, L.D. and Ramani, D.V. (2000). Optimal light integrated and CO_2 concentration for lettuce in ventilated greenhouse, *J.Agric. Engg. Research*, 65(2):129-142.

Feuilloley, P. and Issanchou, G., (2000). Greenhouse covering materials measurement and modeling of thermal properties using hot-bore method and condensation effects, *J.Agric. Engg. Research*, 65(2), 129-142.

Fuchs, M. (1993). Transpiration and foliage temperature in a greenhouse. International Workshop on Cooling Systems for Greenhouses, *Agritech, Tel Aviv*.

Garzoli, K. and Blackwell, J. (1973). The response of greenhouse to high solar radiation and ambient temperature, J.Agric. Engg. Research,. 18, 205-206.

Gastra, P., (1950). Photosynthesis of crop plants as influenced by light, carbon dioxide, temperature, and stomatal deffusion resistance, Meded. Landbouw. *Wageningen*, 59, 1-68.

Ghosal, M. K., Tiwari, G.N. and Srivastava, N.S.L. (2002a). Modeling and experimental validation of greenhouse with evaporative cooling by moving water-film over external shade cloth, *Energy and Buildings*.

Goldsberry, K.L. (1970). Condensation in plastic greenhouse,. *Colorado Flower Growers Asso.* Bull., 244, 1-2.

Hannan, J.J., (1969). Temperature in carnetion flowers, *Colorado Flower Growers' Association Bull.*, 226, 1-3.

Hannan, J.J. and Heins, R.(1975). Effect of plant density on two years of carnetion production,. *Colorado Flower Growers' Association Bull.*, 302, 1-3.

Harmanto, H.J. Tantau and V.M. Salokhe, (2006). Microclimate and Air Exchange Rates in Greenhouses covered with Different Nets in the Humid Tropics. *Biosystems Engineering* 94(2): 239–253.

Harmanto, H.J. Tantau and V. M. Salokhe, (2006). Influence of Insect Screens with Different Mesh Sizes on Ventilation Rate and Microclimate of Greenhouses in the Humid Tropics., *Agr. Engng.* Intl., Vol. VIII. Manuscript BC 05 017.

Harmanto, V.M. Salokhe, M.S. Babel, H.J. Tantau (2005). Water requirement of drip irrigated tomatoes grown in greenhouse in tropical environment,. *Agricultural Water Management*, 2005, Vol. 71, 225–242.

Helmy Y. I.; Singer S.M.; El-Abd S.O. and Abou Hadid A.F. (1996). Environmental modification of sex expression in cucumber plants. Ishs symposium on strategies for market oriented greenhouse production, march 11-15, 1995 Alexandria-Egypt. *Acta Hort.* (434): 361-366.

Holder, R. and Cockskull, K.E. (1990). Effects of humidity on the growth and yield of glasshouse tomatoes, *J. Hort. Sci.*, 65, 31-39.

Jamal K.A. (1994). Greenhouse cooling in hot countries, *Energy*, 19(11): 1187-1192.

Jensen, H.M. and Malter, A.J. (1995). Protected Agriculture – Aglobal review, *World Bank technical paper*, No. 253, p.157.

Kacira, M., T. H. Short and R.R. Stowell. (1998). A CVD evaluation of naturally ventilated, multi-span, sawtooth greenhouses. *Trans of the ASAE* 41(3):833-836.

Kinet, J.M. and M.M. Peet. (1997). Tomato. In The Physiology of Vegetable Crops, ed. H.C. Wien. Commonwealth Agricultural Bureau (CAB) International, Wallingford, UK. 600 p.

Kittas, C., T. Boulard and G. Papadakis. (1997). Natural ventilation of a greenhouse with ridge and side openings: sensitivity to temperature and wind effects. *Trans of the ASAE* 40(2): 415-425.

Lee, I-B. and T. H. Short. (1998). A CFD model of volumetric flow rates for a naturally ventilated, multi-span greenhouse. *ASAE Paper* no. 98-7011, pp 17.

Lee, I-B. L. Okushima, A. Ikeguchi, S. Sase and T.H. Short. (2000). ASAE Paper no. 00-5003, pp 16.

Marcelis L.F.M. (1993). Fruit growth and biomass allocation to the fruit in cucumber. 1. Effect of fruit load and temperature. *Scientia Horticulturae* 54(2) 107-121.

Marcelis L.F.M. (1994b). Fruit shape in cucumber as influenced by position within the plant, fruit load and temperature. *Scientia Horticulturae* 56(4) 299-308.

Misra, R.L. and Pathania, M.S., For north Indian conditions, hi-tech greenhouse and production strategies for cut flowers, *Indian Horticulture*, 9 (April-June, 2000).

Montero, J.I., P. Munoz and A. Anton. 1997. Discharge coefficients of greenhouse windows with insect-proof screens. *Acta Hort.* 443: 71-77.

Mutwiwa, U. N., H. J Tantau and V. M. Salokhe (2006). Response of Tomato Plants to Natural Ventilation and Evaporative Cooling methods. Proceedings of the *International Symposium on Greenhouse Cooling: methods, technologies and plant response,* held at the Hotel Playadulce, Almeria, Spain, from 24th to 27th April 2006. To be published in *Acta Horticulturae.*

Nelson, P.BV., (1984). Greenhouse operation and maintenance, *Reston Pub. Co.*, Vergenia, USA.

Rebuck, S.M., R.A. Aldrich and J.W. White. (1976). Internal curtains for energy conservation in greenhouses. ASAE Paper no. 76-4009, pp 11.

Rosa, R. Silva, A.N. and Miguei, A. (1989). Solar irradiation inside a single span greenhouse, *J. Agric. Engg. Res.*, 43, 221-229.

Rylski, I. and Spigelman, M., (1982). Effect of Different diurnal temerature combination on fruit set of sweet pepper, *Sci. Hort.*, 17:101-106 .

Rylski, I. and Abraham,H., (1987). Tomato fruit development under different temperature regims in net and plastic covered greenhouse, *Hasadeh,* 58, 1277-1265.

Savage, M.J. and Smith, I.E. (1980). The radiation environment of tunnel grown tomatoes: row spacing and pot volume effect, *Crop Prod.*, 9, 223-226.

Saleh, M.M., Abou-Hadid, A.F., El-abd, T. and Shanan, A.S. (1994). Effect of different shading density's on growth and yield of tomato and cucumber plants. *Egypt J. Hort.* (21):1.

Salman, S.R., A.f. Abou-hadid; M.O. Bakry and A.S. El-Beltagy (1991). The effect of plastic mulch on the microclimate of plastic house. *Acta Horticulture*, (287):417-426.

Salman, S.R.; A.f. Abou-Hadid; M.S. El-beltagy and A.S. El-beltagy. Plastic house microclimate condition as affected by low tunnel and plastic mulch.*Egypt. J. Hort.* 19(2):(1992).

Seeman, J., Climate and Climatic Control in the Greenhouse, *Bayer, landw. Verl.* (1957).

Seginer, I. and A. Livne. (1978). Effect of ceiling height on the power requirement of forced ventilation in greenhouses. *Acta Hort.* 87:51-68.

Seginer, I. 1994. Transpirational cooling of a greenhouse crop with partial ground cover. Ag. and for. Meteor. 71:265-81.

Seginer, I. 1997. Alternative design formulas for the ventilation rate of greenhouses. *J. of Ag. Eng. Res.* 68(4): 355-365.

Seginer, I. and M. Tarnopolsky. (1999). Ground treatments to relieve heat stress of a sparse greenhouse canopy. ASAE Paper 99-4124, 16pp.

Seginer, I., D. H. Willits, M. Raviv and M. M. Peet. (2000). Transpirational cooling of greenhouse crops. BARD Final Scientific Report IS-2538-95R, Bet Dagan.

Short, T.H. and W.L. Bauerle. 1977. A double-plastic heat conservation system for glass greenhouses. ASAE Paper no. 77 4528, pp 10.

Simpkins, J.C., D.R. Mears and W.J. Roberts. (1975). Reducing heat losses in polyethylene covered greenhouses. ASAE Paper no. 75-4022, pp 14.

Sirjacobs, M. (1989). Greenhouse in egypt, protected cultivation in the mediterranean climate, *FAO,* Rome, Italy.

Sirohi, P.S. and Behera, T.K., (2000). Protected cultivation and seed production in vegetables, *Indian Horticulture*, p.23.

Smith, I.E., Savage, M.J. and Mills, P., (1986). Shading effects on greenhouse tomatoes and cucumber, *Acta Horticulture,* 2 (184), 491-500.

Srivastava, N.S.L., Din, M. and Tiwari, G.N. (2000). Performance evaluation of distillation cum greenhouse for warm and hot climate, *Desalination,*128, 67.

Sutar, R.F. and Tiwari, G.N. (1996). Reduction of temperatureinside a greenhouse, *Energy,* 21, 61-65.

Teitel, M., Barak, M., Berlinger, M.J., Mordechai and Sara, L. (1999). Insect proof screens in greenhouses: their effect on roof ventilation and insect penetration. *Acta Horticulture*, 507: 25-34.

Teitel, M. and J. Tanny. (1999). Natural ventilation of greenhouses: experiments and model. Ag. and For. Meteor. 96:59-70.

Teitel, M., Segal, I., Shklyar, A. and Barak, M., (1999). A comparison between pipe and air heating methods for greenhouse, *J. Agric. Engg. Research,* 72(3): 259-273.

Tiwari, G.N., and Goyal R.K., (1998). Greenhouse Technology, *Narosa Publishing House,* New Delhi, 252-311.

Tiwari, G.N., (2003). Greenhouse technology for controlled environment, *Narosa Publishing House*, New Delhi.

Virhammar, K. (1982). Plastic greenhouses for warm climates. *FAO agricultural services bulletin* 48. Food and Agriculture Organization of the United Nations. Rome.

Verlodt H., (1990). Greenhouse in Cyprus, protected cultivation in the Meditarranean climate, *FAO*, Rome, Italy, .

Walker, J.N., (1987). Predicting temperature in ventilated greenhouse, *trans. ASAE*, 8(3)445-448.

Walker, J.N. and Duncun, G.A., Greenhouse location and Orientation, *Kentucky Univ. Misc. Publication*, 397 (1971).

Walker, P.N. (1979). Greenhouse surface heating with power plant cooling water: heat transfer characteristics. *Trans of the ASAE*. 22(6):1370-1374.

Wang, S. and J. Deltour. (1999a.). Airflow patterns and associated Ventilation function in large-scale multi-span greenhouses. *Trans of the ASAE* 42(5):1409-1414.

Wang, S and J. Deltour. (1999b). Lee-side ventilation-induced air movement in a large-scale multi-span greenhouse. *J. of Ag Eng. Res*. 74:103-110.

Wang, S. and Boulard, T. (200a). *Measurment* and prediction of solar radiation distribution in full-scale greenhouse tunnels, *Agronomie,* 20, 41-50.

Wang, S. and Boulard, T. (2000b). Predicting the microclimate in a naturally ventilated greenhouse in a Mediterranean climate, *J. Agric. Research,*75(1): 27-38.

Willits, D.H. (1999a). Constraints and limitations in greenhouse cooling: Challenges for the next decade. In: Proceedings of the International Conference on Greenhouse Techniques Towards the 3rd Millennium, 6-8 Sept, Haifa, Israel, eds: M. Teitel and B.J. Bailey. *Acta Hort*. 534.

Willits, D.H. (1999b). The effect of canopy density, ventilation rate and evaporative cooling on the transpiration of a greenhouse tomato crop. *ASAE Paper* no. 99-4227:18 .

Willits, D. H. (2000a). Design of environmental control systems for greenhouses: A comparison of four engineering standards plus input from the industry. ASAE Paper no. 00-4060, pp 18 .

Willits, D. H. (2000b). The effect of ventilation rate, evaporative cooling, shading and mixing fans on air and leaf temperatures in a greenhouse tomato crop. ASAE Paper no. 00-4058, pp 18.

Willits, D.H. and Bailey. D.A. (2000). The effect of night temperature on chrysanthemum flowering: heat tolerant vs. heat sensitive cultivars. *Scientia Hort*. 83(3):325-330.

Willits, D.H., Ahmad I, and Peet M.M. (1991). A model for greenhouse cooling. ASAE paper no. 91-4041, pp 23.

Willits, D.H. and Peet, M.M. (2000). Intermittent application of water to an externally mounted greenhouse shade cloth to modify the cooling performance, *Transactions of the ASAE*, 43(5):1247-1252.

Winspear, K.W. and Bailey B.J. (1978). Thermal screens for greenhouse energy effectiveness. *Acta Horticulturae* 87: 111-118.

Wright, W.J. (1917). Greenhouse construction and equipment, *Prange Judd. Co.,* New York.

Zaiqiang Yang, Yushan Li, Ping Li, Fangmin Zhang & Ben W. Thomas (2016). Effect of difference between day and night temperature on tomato (*Lycopersicon esculentum* Mill.) root activity and low molecular weight organic acid secretion, Pages 423-431 | Received 26 Jun 2016, Accepted 11 Aug 2016, Published Online.

Index

Printed and bound by CPI Group (UK) Ltd, Croydon, CR0 4YY

17/10/2024

01775682-0004